SuperAnimals and Their Unusual Careers

Westminster Press Books
by
VIRGINIA PHELPS CLEMENS

SuperAnimals and Their Unusual Careers
A Horse in Your Backyard?

SuperAnimals AND Their Unusual Careers

By

VIRGINIA PHELPS CLEMENS

THE WESTMINSTER PRESS
Philadelphia

BOOK DESIGN BY DOROTHY ALDEN SMITH

First edition

Published by The Westminster Press®
Philadelphia, Pennsylvania

PRINTED IN THE UNITED STATES OF AMERICA

9 8 7 6 5 4 3 2 1

PICTURE CREDITS: Animal Behavior Enterprises, Inc., pp. 47, 55, 59, 61, 74; Everett Hayes, p. 173; Lincoln-Mercury Division, Ford Motor Company, p. 73; Mulberry Square Productions, Inc., pp. 115, 116; 9-Lives Cat Food, p. 72; The Philadelphia Inquirer, pp. 56, 121; Ralston Purina Company, p. 65; June Stefanelli, p. 19; Wrather Corporation, pp. 111, 113.

Portions of the chapter "An Apple for the Teacher" have appeared in *Horse of Course!* magazine. © 1979 Derbyshire Publishing Company, Inc., Temple, N.H. 03084.

Library of Congress Cataloging in Publication Data

Clemens, Virginia Phelps.
SuperAnimals and their unusual careers.

Includes index.
SUMMARY: Text and photographs describe the different kinds of jobs performed by various animals including German shepherds that sniff out bombs, ponies that help disabled riders, and hearing-ear dogs.
1. Working animals—Juvenile literature.
1. Working animals. 2. Animals I. Title.
SF172.C57 636.08'86 79-10932
ISBN 0-664-32649-8

For all my animal friends, past and present,
with love

I wish to thank the following people for helping me gather the information and photographs necessary to complete this book:

Marion Bailey
Moe DiSesso
Starr Hayes
Elaine Lehr
Carl Miller
David Premack
June Stefanelli
Bannie Stewart

Contents

Introduction

Many pets and other animals are working every day, all around us, without our even knowing about them. They are more than just companions. They are partners, teachers, entertainers, and lifesavers. They are doing their jobs faithfully and without any complaining, whether they are in front of a television camera, hundreds of feet below sea level, or in a local gymnasium.

Monkeys are taught to do many tricks, but did you know that they are also learning how to "talk"?

Did you know that dolphins are used to rescue lost divers?

Or that horses are being used to help disabled children?

Or that dogs have been trained to find bombs?

The dolphins, dogs, horses, cats, monkeys, and other animals described in this book deserve recognition for the jobs they are doing. Their stories and information about them in the following chapters will help you become acquainted with the skills of animals. People depend on little-known animal services every day to make their own lives and occupations easier.

SuperAnimals offer very special talents. Our lives would be poorer without them.

1
The Nose Knows Best

"There's a bomb on Continental Flight 608," said the harsh-sounding voice on the other end of the line. Then, click, the caller hung up.

The airport official replaced his receiver and paused for a second before making a call to Officer Jim Manning.

"Jim, I just received an anonymous phone call. We have a bomb scare on Continental 608," he said quickly. "Get Alex down to the end of Runway 6. I'll call the pilot of the plane and tell him to turn around and meet us there."

Runway 6 was the farthest runway from the terminal building. By the time the airport official got there, Continental Flight 608 had already arrived and opened its doors. The passengers were beginning to file off, carrying their coats and any small packages they had brought on the plane themselves. Their luggage was being unloaded from the cargo hold by the loading crew.

Officer Manning stood patiently nearby with Alex, a

large black-and-tan German shepherd, sitting at his left side.

Finally all the passengers had left the plane.

"OK. I'm taking Alex inside now. Please keep everyone out. I'll be ready to search the luggage in a few minutes," Officer Manning told the airport official. "Have the bomb crew ready in case he finds something."

The young man quickly climbed the portable stairs and entered the plane. Alex trotted happily at his heel. He knew he was going to be able to play the game soon.

Inside the plane, Officer Manning unsnapped the leash and said, "O.K., Alex, find the bomb."

Eagerly, the dog trotted down the aisle, sniffing at each seat. Tail wagging, ears pricked forward, and nose drawing in the various smells, Alex searched for "that scent." This was a game he loved to play. Unfortunately, he couldn't always find "the scent." Sometimes it wasn't where Officer Manning sent him to hunt for it.

After making a complete tour of the plane, including the cockpit, galley (kitchen), and rest rooms, the shepherd trotted back to Officer Manning and sat down in front of him. Disappointment showed in every line of his body from his drooping tail to his limp ears and low-hung head.

"That's all right, Alex. We're not finished yet."

Officer Manning patted him sympathetically and snapped the leash onto his collar. Then together they searched the passenger cabin again. This time Officer Manning went with Alex, pointed to various areas, and said, "Is it there? Find it! Find the bomb, Alex."

But they found nothing unusual inside the plane.

Outside again, the partners walked over to where the luggage had been placed in a long line down the runway, five feet between each piece.

"Everyone stand back, please," Officer Manning said to the small group of men nearby.

"Give him room to work."

Turning to the dog sitting at his side, he unsnapped the leash again and commanded, "Alex, find the bomb!"

Alex bounded forward, ears erect and nose already exploring the air currents. At the first piece of luggage he slowed his pace, gave a sniff, and continued on down the line, barely turning his head at most of the suitcases.

Suddenly, he stopped in mid-stride and turned toward a dark-brown briefcase.

"Alex, heel!"

The dog whirled and returned quickly to his partner. Disgust showed all over his face. That was it. He knew

[PAGES 14–15]
A drug-sniffing dog finds two suspicious briefcases. He decides the drug is in only one case, points to the hiding place, then carefully watches his handler retrieve the hidden drug

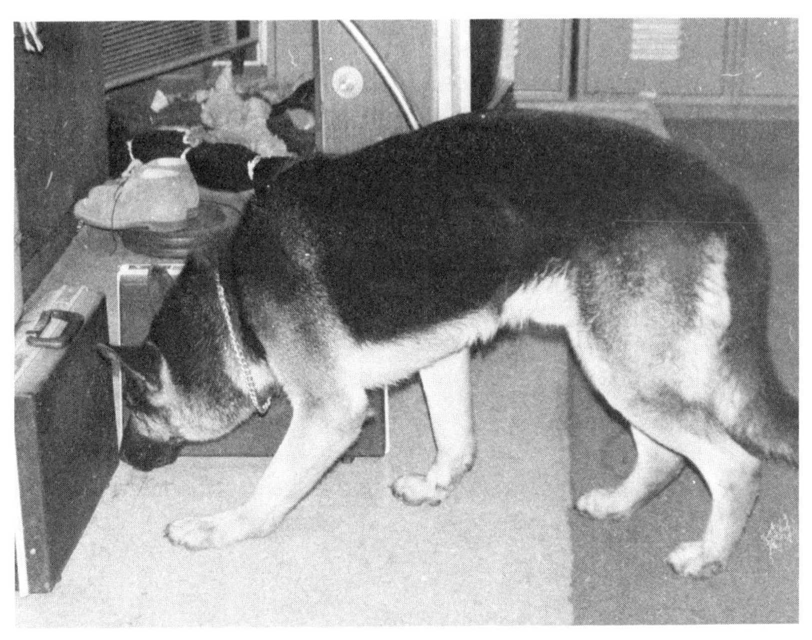

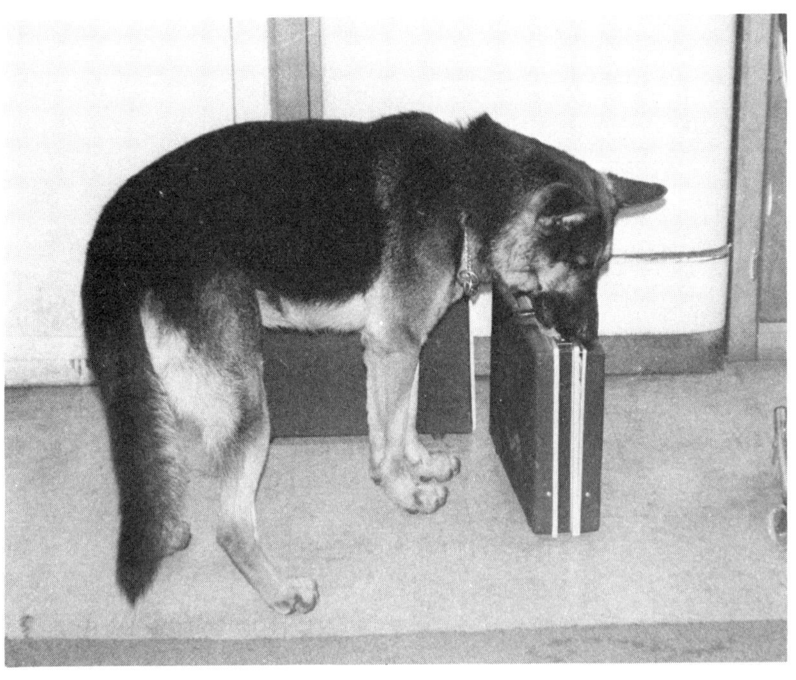

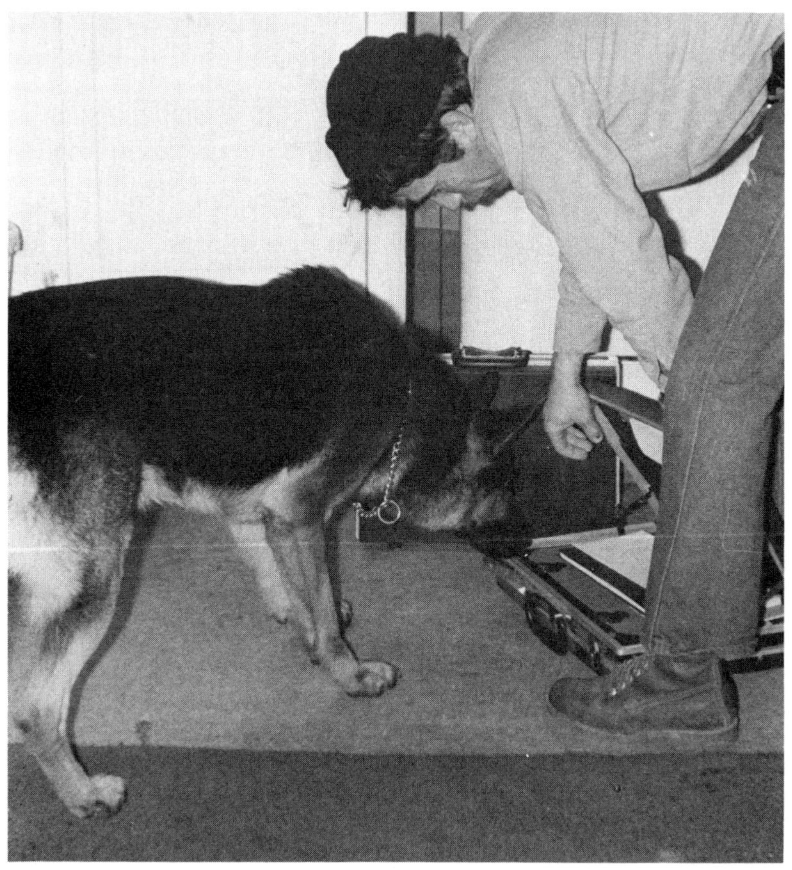

"the scent" was in there and he wanted to get at it.

"That's it," said Officer Manning to the other men standing behind him. "The brown briefcase."

"Are you sure?" asked one of them in disbelief.

"Yes, let the bomb crew handle it now." He bent down and patted the dog heartily.

"Good boy, Alex. I knew you could find it. Good boy."

In spite of his disappointment at not being able to retrieve "the scent" and play with it, Alex's eyes brightened and his body wriggled with joy at his partner's praise.

As the bomb crew moved in to take possession of the suspected case, Officer Manning walked back toward the terminal building. Alex followed at his side, head held high and tail wagging happily with every step. He had performed his job perfectly, saving hours and possibly lives with his nose.

* * *

All over the world dogs like Alex are using their keen sense of smell to search for explosives and narcotics (marijuana, hashish, heroin, opium, and cocaine). They are on duty in airports, at seaports, and at border crossings. They have been sent to inspect post offices, warehouses, banks, factories, schools, private homes—in fact, anywhere there has been a bomb threat or the possibility of drugs or narcotics being stored illegally.

Drug- and bomb-sniffing dogs inspect everything, including trash cans, to find illegal goods

Most of the dogs used for this work are German shepherds, although other large breeds of sporting or working dogs—Labrador retrievers, Weimaraners, collies, Irish setters, Shetland sheepdogs, and even a combination of them (mutts)—are used for scent work, too. Small dogs such as beagles, fox terriers, and cocker spaniels have a good sense of smell, too, but they are not large enough for patrol work.

German shepherds have stamina, do well in both hot

and cold climates, and have the intelligence necessary for patrol and scent work. They usually stand about 23 inches at the shoulder and weigh approximately 80 pounds. It is the handler's responsibility to keep his dog in good physical condition without any excess weight, because too much fat will slow him down, especially in hot weather.

Male dogs are preferred because of their steadier temperament and larger, more impressive size. They should be between one and three years old, because younger dogs are too immature and not yet fully grown, and older dogs have too few years of service left. The average life span of a dog is approximately ten years. If a six-year-old dog was accepted for training, he would probably be able to work for only a couple of years.

The most important requirement for a possible drug or bomb detector is for the dog to want to retrieve. The dog must be alert, show aggressiveness toward a thrown object, and have a curious, bold, and willing disposition. He must also have a strong power of concentration so he is not distracted from his job by other animals, loud noises, or curious bystanders.

The United States Customs Service, which employs over fifty dogs for bomb and narcotics detection, has found that only one out of every 130 dogs is eligible for scent work. Many flunk the physical examination during which they have their blood, heart, ears, eyes, and lungs tested and their hips X-rayed for hip dysplasia.

During obedience training, drug-sniffing dogs learn to climb ladders and to jump various obstacles

Others are just unreliable, lazy, too scared of people or strange surroundings, gun-shy (afraid of loud noises), or lacking the desire to find "the scent."

Many of the dogs trained for scent work have been donated because they grew too large or too uncontrollable for their owners. People often turn to local police departments or the Customs Service when their cute little puppy grows up into a rowdy, oversized monster jumping up on everyone, charging through the house, and knocking down children. Some of these dogs are even rescued at the last minute from the gas chambers of humane shelters.

Dutchess, a German shepherd who bit a mailman, was on her way to being put to sleep in a Pennsylvania shelter. However, she was bought and trained to find narcotics instead.

The Philadelphia Police Department has found out that dogs picked up at humane shelters are usually very "street wise." They are well acquainted with cars, people, and the noise of a city, because they have been roaming the streets as strays. Usually these "alley shepherds" have worms and fleas, are skinny and dirty, and have been mistreated, but they respond well to love, care, and affection. They prove very successful at scent work because the noise and confusion of a city rarely upsets them. A pampered, sheltered dog always kept in a yard, house, or kennel may not do so well. Typical city surroundings—a crowded subway, bustling airport, or dark alley crawling with rats and resembling an obstacle course with its barrels of garbage and mounds of trash—can be very frightening to a house pet.

Once accepted into the program, the dog is introduced to his handler. During the first week, dogs may be switched around with different officers until the trainer feels every pair is a good working team with matching personalities. The partnership between a handler and his dog is very important. Some dogs need sterner, more physical handlers, while others need the light touch of a gentler, more sensitive person. Once a team is established they will be together practically day and night for the rest of the dog's working life.

After a day's work the dog is not put into a cage or a kennel until the next day. He goes home with his partner and becomes one of the family. He plays with the children, stretches out under the table at mealtimes, and works the same shifts as his human partner, with the same vacations and days off.

The dogs are first trained for patrol work. This takes from eleven to fourteen weeks, depending on the trainer. They learn basic obedience: to come, sit, heel, and lie down—on and off lead (leash), by both voice and hand signals. They learn to track (find objects and people by following a smell), and to catch a lawbreaker by grabbing on to an arm, ankle, or leg and holding him until the handler arrives and gives the signal to let go. They also learn to protect their handlers. Often, a dog who has been brainwashed from puppyhood never to bite has to learn from watching the other dogs how to attack someone who is threatening to hurt his partner.

Dogs have long memories but short attention spans, so instruction periods are short, with lessons repeated over and over again. Dogs can be forced to learn, but those who want to learn are much more successful pupils. Training periods are serious but enjoyable. A happy dog is a much more reliable worker, so rewards, a few words of praise or a couple of hearty pats on the shoulder, are handed out whenever they are earned.

During the course each handler should be learning

about his own dog—recognizing what the dog can and cannot do and becoming more aware of its moods and feelings. Some dogs pick up commands quickly and follow them without hesitation, while others are slower but can think on their own without always depending on a handler's commands to guide them.

Any dog that has a major problem, such as turning vicious and biting everyone, is dropped from the program, because there isn't enough time to correct it. Minor problems, such as paying too much attention to prowling cats or another dog, are easily dealt with during class.

At this point in their training, some dogs show that they will make excellent patrol dogs but have no real desire to hunt for and retrieve an object. These dogs never start the next training period, but remain on patrol duty.

The next six to nine weeks are spent on learning to find narcotics or explosives. Some departments use only dogs that have been tested in a year or two of patrol work. Others take their dogs from the first training period and put them right into the second. The best graduates are those that enjoy scent work and have a true instinct for finding things on their own.

Dogs are trained to find either narcotics or explosives. They are not trained for both, because this requires two different kinds of training. However, both kinds of training are built on stimulus, response, and reward. When the dog responds to a smell (stimulus) and retrieves or signals that he sees the correct object, he is rewarded with praise and a pat or food tidbit.

On the first day of scent training, the dogs are introduced to articles heavily scented with marijuana, gunpowder, or whatever scent they are being trained to find. First, the articles are thrown for the dog to retrieve. These may be a bundle of rags, a plastic bag, a wooden dummy, or a hollow ball containing some of "the scent."

Then the object is placed in sight, but at a distance from the dog, so he learns to retrieve objects already in place. When he learns to pick out the object with "the scent" from other objects without "the scent" and bring it to his handler, the objects are hidden. Now the dog must find them by using only his nose. The dog must also learn to find the object even though "the scent" is disguised by talcum powder, perfume, surgical soap, or incense.

These lessons don't always take place in the training center. The dogs are taken as often as possible to houses, banks, post offices, airports, and other buildings so they can become accustomed to working in unfamiliar locations.

Dogs trained to find narcotics will go right after them, pawing at suitcases, scratching at closet doors and furniture, or tearing apart trash cans and boxes to get at them. When searching a car they will go from front seat to back, even crawling underneath the car if they get a whiff of their target. Wherever it is hidden

—in the trunk, behind a headlight, or in a hubcap, they will find it.

Care must be taken with dogs who are trained to find heroin, because pure heroin will burn holes in the membranes of a dog's nose if the dog touches it. The dogs could also become addicted to it by smelling it too often. Blood tests are given regularly to prevent this. If there is a chance that a dog is "hooked" on heroin, he is given a vacation until cured.

Dogs trained to find explosives, such as gunpowder, can start their training in the same way as narcotics-sniffing dogs. However, those trained to find dynamite must learn their job with only the paper in which dynamite has been wrapped. This explosive is not safe after being stored for any period of time and may explode on its own without warning.

Instead of retrieving the explosives, the dogs must be trained to signal their discovery. Any shaking or pressure could cause it to explode. Some are taught to sit as soon as they smell an explosive and wait for a reward—either a ball to play with or a piece of food. Others are called to a heel position as soon as their carefully watching handler sees that they have caught "the scent" in a particular area. This is why a handler must know his dog's reactions to different situations. Since the dog cannot talk, the handler must be able to recognize and understand the signals the dog gives with his body—the wagging of his tail, whining, ears laid back close to his head, or a sudden change in direction.

A dog's nose is 30,000 times more sensitive than a person's nose. A dog's sense of smell is so sharp that dogs have found narcotics in light switches, in the bottom of potted plants, in tins of dusting powder, radios, beer steins, and boxes of candy. They can even point out a person who has recently handled a drug or an explosive.

The "drug dogs" are also being used in schools to sniff out marijuana, cocaine, and heroin hidden in lockers, boys' and girls' rooms, desks, and other hiding places. It is hoped that they will reduce drug possession in the schools by 80 to 90 percent.

Trep, who works for the Miami, Florida, police department, has sniffed out over $63 million worth of illegal narcotics by himself so far. During a demonstration he surprised everyone by finding eleven packets of drugs on a subject—when only ten had been hidden.

One of the dogs with the Philadelphia Police Department found marijuana hidden in a sealed lead container. Yogi, a yellow Labrador retriever with the London, England, Metropolitan Police, is even able to point out narcotics hidden under water.

Dogs trained to find explosives have been just as successful. In Oakland, California, an explosives-detecting canine found a bomb in the ceiling—several feet above his head. Another California dog discovered a letter bomb addressed to the mayor of Los Angeles.

A dozen German shepherds have been recruited to help protect the President. Every time President

Carter holds a formal news conference, the room in which he meets the reporters is checked for explosives by Secret Service agents and their bomb-sniffing dogs. Unfortunately, the videotape in television cameras seems to smell just like bombs, so the dogs will often get excited when a camera is rolled into the room.

Dogs have proved quite valuable in bomb and narcotics detection because of their ability to perform their job many times faster than people. One dog and one handler can check from 1,000 to 1,800 packages in thirty minutes—a task that would take one person several days. One dog can inspect a car in about two minutes, which is ten times faster than a man can inspect it.

Yogi, the London retriever, participated in over 2,000 drug searches, resulting in more than 1,200 arrests and convictions during his six years of service with the London Metropolitan Police drug squad.

Smokey, of the United States Customs Service, checked more than four and one half million vehicles, mail packages, and cargo units in just one year. During this time he uncovered large quantities of opium, heroin, and pills, and nearly eight tons of marijuana, all of which had a street value estimated at over $4.6 million.

One of the Philadelphia Police Department's shepherds stationed at Philadelphia International Airport during a seven to eight month period found over 400 pounds of marijuana just by sitting next to the luggage belt and sniffing the packages and suitcases as they went by him. Another one of Philadelphia's shepherds proved that he was always on the job even when off duty. Let loose for some exercise in the Philadelphia airport's auto pound one morning, he returned with a bundle of counterfeit money wrapped in some rags.

In 1972, Brandy, a bomb-sniffing German shepherd on duty at New York City's John F. Kennedy International Airport, found a bomb on a TWA jet just twelve minutes before it was set to go off. This incident helped start a federal program that now has "bomb dog" squads all over the country. Stationed at various airports, they are within thirty minutes of any commercial airliner flying over the continental United States. The Law Enforcement Assistance Administration (LEAA) pays all expenses and provides dogs for local policemen, sheriff's officers, or airport security men sent to Lackland Air Force Base in Texas for training.

Narcotics- and bomb-detecting dogs have proved to be accurate 80 to 90 percent of the time, depending on the people involved in the search. Occasionally, curious bystanders get in the way of a working dog. Outside evaluators of the LEAA's special dogs found them able to locate hidden explosives 96.6 percent of the time. The average time was sixteen minutes, but some bombs were found in only thirty seconds.

Just like people, dogs may have good days and bad days. A cold that interferes with his sense of smell, or even a dull, sluggish feeling, can keep a dog from working at his best. If a dog is sent out on a job and doesn't find anything, he may be tested to see if he is working all right. After the dog is taken from the area, another officer will hide a dummy—a demonstration explosive or drug. The dog will be brought back and commanded to search again. If he finds the dummy, his handler will know that he is feeling all right and could find other drugs or bombs if they were hidden there, too. It is harder for the dog to find the dummy, because any narcotic or explosive allowed to sit for a while—six to seven hours—has a stronger odor than one placed in the room within the past few minutes.

Even after the training sessions, dummy explosives and narcotics are used with the dogs. If a dog is constantly sent out on false alarms, never finding anything because nothing is there, he could become discouraged and doubt his ability to find "the scent." To prevent this, a handler will have another officer hide a dummy whenever his dog doesn't find anything on a search. Then the dog is commanded to "Fetch the pot" or "Find the bomb." When he finds it, he is allowed to play with it as he was able to do during training. This is his reward and one he looks forward to when he finds "the scent."

The dogs are also given refresher courses. Working on the street does not usually provide enough action to keep a dog's nose sharp and his body in shape, so he is brought back to the training center. There is never any failure here, and they know they will have to work hard. Constant success during planned searches at the center gives them confidence in their own ability again.

Dogs on duty at the airport are frequently tested with dummies there, too. These exercises are fun for the dogs, but also a valuable way to keep them alert.

Scent-working dogs are retired when they are between eight and thirteen years old, depending on their health and ability to work. Usually they live out their last years with their handler's family. However, if an old partner cannot accept his handler's new canine partner, a new home is found for him with another family. The dogs are never just put to sleep or shuffled off to the SPCA. Law enforcement agencies' most valuable partners in crime detection are rewarded, not rejected, after their years of service.

Dogs have been trained to find other objects with their noses, too. They are used to find leaks in gas lines; gypsy moth clusters that can ruin an entire forest in one season; termite nests that can weaken the structure of a wooden house; truffles, a type of underground mushroom that is used in France by gourmet cooks; and even people buried in avalanches in Switzerland.

According to an old wives' tale, a cold, wet nose is supposed to be a sign that a dog is healthy. Today, dogs' cold, wet noses are saving hundreds of lives and preventing millions of dollars in property damage every year. Surely this is a healthy sign, too.

2
An Apple for the Teacher

Up, down, up, down.

The ponies trotted slowly around the ring. Their riders tried to post in time with their steps.

"Hands down. Grip with your knees," called out Mrs. Bannie Stewart, the riding director.

The children worked hard to pick up the rhythm, pushing themselves up out of the saddles and then sitting down again. Up, down, up, down. A couple of them were posting with only an occasional off beat, but the rest still bounced in the saddles.

The ponies were led by teen-age volunteers, and side helpers—also young volunteers—ran along on each side of the ponies. The side helpers kept an eye and occasionally a hand on the riders to steady their balance.

The ponies followed the track around the edge of the ring. Up, down, up, down. April slowed her pace even more because she could feel her rider wobbling. She didn't want Debbie to fall off.

April was a small black pony with a white star on her

forehead. She was only four years old, which was unusually young for this type of work. Her companions in the ring, Pixie, Smoky, Twiggy, Buttercup, and Velvet, were all older. In fact, Velvet, a shiny black pony who looked just like April, was April's mother. Velvet had arrived at the barn in foal and surprised everyone one bright spring morning with a tiny black filly, who was named April after the month in which she was born.

"Walk, please," said Mrs. Stewart, who stood in the middle of the ring.

April gradually slowed to a walk, not minding the uneven jerking on her mouth as the little girl pulled back sharply on the reins. April knew that Debbie was physically disabled, that she didn't mean to hurt her.

Debbie had had a stroke when she was only five years old, leaving her practically blind and with poor control of her arms and legs. Her movements were jerky, and often her arms and legs didn't do what she wanted them to do. When she was riding April, she tried harder than ever to control her hands so she didn't pull unnecessarily on the reins and hurt April's mouth. Debbie loved horseback riding, and especially April. Riding was one activity she could do, and she looked forward to her weekly lessons.

"All right, let's line up for our exercises," said Mrs. Stewart.

The riders turned their ponies and walked them to one end of the ring. The ponies lined up facing Mrs.

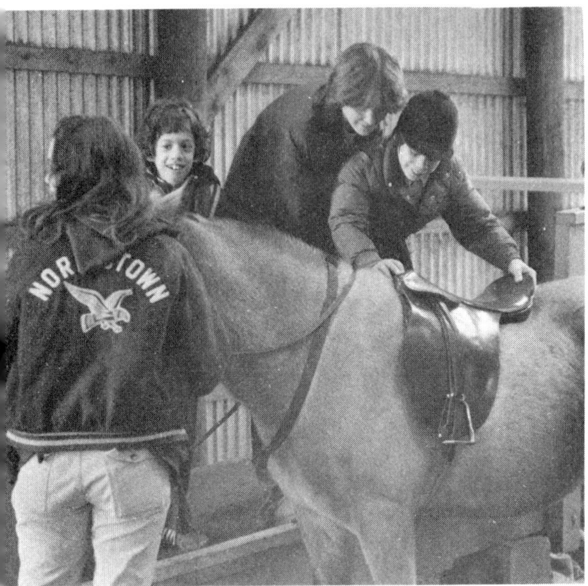

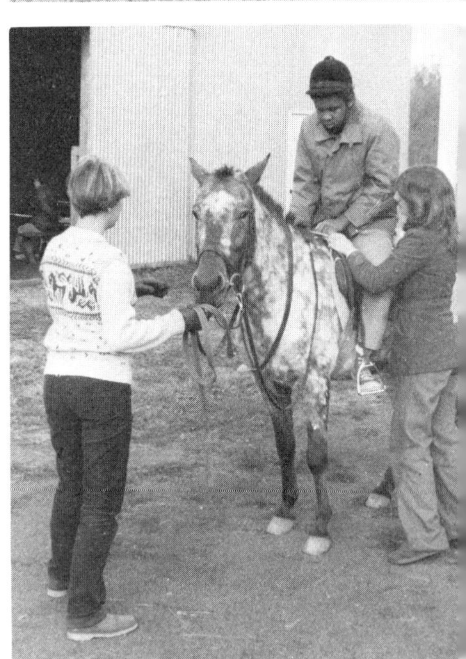

Horseback riding is fun for disabled children. Some of the children must mount from a ramp and some need neck straps and a safety stirrup so they can ride

Stewart and stood quietly side by side.

Following Mrs. Stewart's instructions, the children leaned forward and touched their ponies' ears, smacked their knees with both hands, and then raised their arms as high as they could before reaching again for the ponies' ears. Sometimes the teen-agers at their sides had to help them. The riders also practiced standing up in their stirrups and then settling back down into the saddle. This helped to strengthen their legs for posting when the ponies trotted.

"Don't plop down in the saddle like a sack of potatoes," warned Mrs. Stewart. "It will hurt your pony's back."

April looked sideways to the pony at the other end of the line. Pixie was a little chestnut pony, about eight years old. Her father was a quarter horse and her mother had been someone's family pet.

Standing next to her was a buckskin pony—Smoky. He was in his twenties, still healthy, but not fancy-looking. He had been outgrown by the children in two different families. The last owner, knowing how good Smoky was with children, had donated him to the riding program for disabled children.

Twiggy, a small horse, was a teen-ager, fifteen to be exact, and had been a show horse when she was younger. A flashy bay (brown coat with black mane and tail), Twiggy was half thoroughbred and half Morgan. Her previous owner had won several ribbons and trophies on Twiggy's back, but eventually he had grown too big for her.

Velvet was next to Twiggy, and Buttercup, a round little pony, was between her and April. Buttercup was golden brown in color and also a teen-ager. She was so wide, almost wider than she was tall. No one ever felt as if they were falling off her. Cutting down on her feed and giving her more exercise did little to reduce her size. She almost seemed to put on weight through the air she breathed.

Just like April's rider, Debbie, all the other riders were disabled too. One was completely blind, another had cerebral palsy resulting from brain damage at birth, while the other three had been physically disabled by multiple sclerosis—a disease that attacks the central nervous system.

"O.K., let's spread out and take turns steering the ponies in and out this line of barrels," Mrs. Stewart said, pointing to half a dozen barrels set up in a line down one side of the ring.

"Kick your ponies and wake them up. You all look like you're falling asleep."

This last remark made all the children giggle as they tried to urge their mounts to move a little faster.

"April won't go any faster," complained Debbie.

"Well, no wonder," explained Mrs. Stewart. "You're kicking her and telling her to move on, but pulling back on the reins at the same time and telling her to stop."

"Oh, I didn't mean to," said Debbie as she guiltily loosened up on the reins. "Sorry, April."

April stretched her neck and immediately stepped up her pace. She knew that Debbie hadn't meant to confuse her, but sometimes it was hard to tell what she wanted.

April is a full-time riding teacher for disabled children at Sebastian Riding Associates, Inc., in Wynnewood, Pennsylvania. She is only one of many horses and ponies in stables all over the country helping to teach mentally retarded and physically disabled people how to ride.

* * *

Horseback riding for the disabled, also called therapeutic riding, is one of the most helpful forms of recreational therapy, particularly for children. Horseback riding strengthens and relaxes muscles, and it improves posture, balance, and coordination.

Learning to ride is also a tremendous confidence builder. Most disabled children have to depend on others to help them, but on the back of a horse they find themselves in control for the first time. Up on a horse they can look down at people instead of always looking up from a wheelchair.

Other children can run and play in many activities—jumping rope, playing hide-and-seek, throwing and catching a ball—while a disabled child must sit and watch. Horseback riding is one of the few physical activities they can take part in, too.

Riding programs for the disabled were started over twenty years ago in England and other European countries. There are now about two hundred centers overseas.

Therapeutic riding in the United States is fairly new. Only in recent years have these programs been formed, but now there are over fifty-five centers scattered across the country. Illinois and Michigan have the most of any states. The largest is sponsored by the Cheff Center for the Handicapped in Augusta, Michigan. It was established in 1968, and by 1976 there were 455 students enrolled in its classes.

The North American Riding for the Handicapped Association (NARHA) was formed in 1969. It is a group of about forty organizations that offer riding lessons for the disabled. The Association encourages horseback riding as therapeutic recreation. They inspect Riding for the Handicapped programs throughout North America to make sure that their safety standards and teaching methods are being followed.

The horses are very important in this type of riding program. First they are thoroughly checked by a veterinarian to make sure they are healthy. Then, whether the horse is bought or donated, it is usually kept at the barn for a trial period of two to four weeks. During this period it is tested to see if it can be trusted with disabled children.

The most important requirement for acceptance of an animal is temperament. The ponies and horses in this program are gentle, quiet, and willing to work. New and strange objects do not excite them, and they are able to work with people and other horses all around them. A rider whose signals are not always very clear does not upset them. If a horse kicks, bites, shies, or bucks, it is immediately rejected because of the danger of serious injury to students.

Size is also considered. For children, the smaller ponies are better. For adults, a small horse up to 15 hands (five feet at the withers, or base of the neck) is best. Students are less likely to be afraid of a smaller animal. They feel more secure if they are not too far from the ground. Also a small pony makes it easier for the volunteer side helpers to balance an unsteady rider in the saddle. They don't have to reach up so high to hold the child in place.

The gait of these horses and ponies is important, too. They must move easily, freely and smoothly, because no rider enjoys riding a horse that jolts them with every step. A horse must have three distinct gaits—walk, trot, and canter—so the students can learn to ride in each one. It is very difficult for a beginner, especially one who is disabled, to learn to post on a mount that has an uneven trot, or to try and sit to a rough canter. The animals chosen for this program are also very surefooted, because stumbling can cause an unsteady rider to fall off.

Age, sex, and looks usually are of little concern in considering a potential mount for this program. Naturally, a very young, untrained horse or one that is too old to be ridden much longer is not acceptable. But horses that range from a trained, quiet three-year-old to one in its twenties can be perfectly suitable mounts, depending on the rest of their qualities.

Either mares (females) or geldings (neutered males) are used with disabled riders. Many horsemen and horsewomen feel that geldings are more reliable and even-tempered, but mares seem to be more sensitive to a rider's special needs. They even appear to take better care of their riders.

Good conformation or shape is nice to find, but beautiful, purebred horses and ponies are usually too spirited for this program. All sizes, shapes, and colors are eligible if the horse is healthy. Big ears, narrow chests, knobby knees, and scrawny necks are not even noticed if the horse is sound and has a steady, gentle personality.

After a horse or pony is accepted for a therapeutic riding program, he enters a two-week training period. During this time he is trained to accept and, if possible, taught to take care of his riders. The trainer doing this job must be knowledgeable and patient. A trainer who takes his anger out on his horses and who is too quick to hit a horse that doesn't understand what is expected of him, will produce a nervous mount, unsatisfactory for this program.

Horses and ponies have excellent memories but very poor reasoning powers. They remember bad training

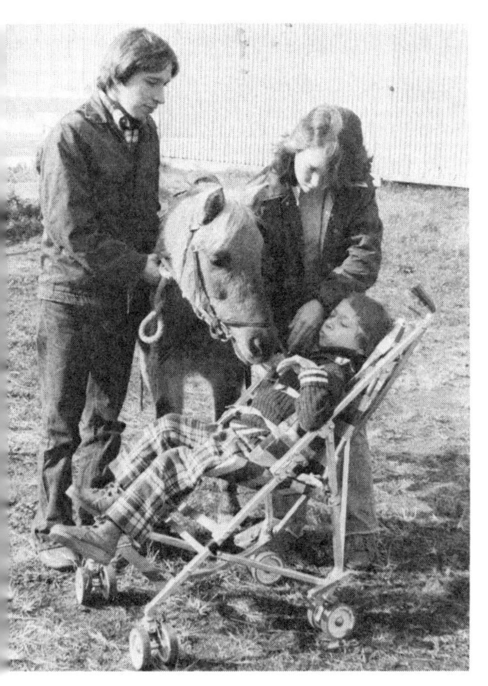

Teen-age volunteers help horse and rider to understand each other and have fun together

as well as good training, and do not realize the difference between the two. A training system of reward and punishment usually has excellent results. Praising and petting a horse that stands quietly while being mounted will help him to know what is expected of him. A good jerk on the reins of one that continually puts his head down to eat grass while being ridden will let him know that this is a bad habit and one that will be punished every time.

The horses and ponies in a therapeutic riding program must learn to stand quietly alongside a wheelchair, and to ignore crutches and walkers. Often a person must use these aids to approach a mount. Reward and praise is also used to teach the horse not to fear a leg brace clanking on a stirrup or a cane bumping against his side. If he is allowed to sniff and inspect strange-looking objects, he will soon learn that they won't hurt him. He should even be allowed to knock over the barrels, poles, and buckets used in the games during the lessons so he won't be scared if this does happen when he has a child on his back.

New horses must also become acquainted with a mounting ramp. A mounting ramp lets even the most severely disabled person mount a horse quite easily. A wheelchair can be rolled up the incline to the top, which is about the same height as the horses' backs. The horses have to learn to stand quietly next to the ramp, sometimes for several minutes, because it is often very difficult for a disabled rider to mount.

The ponies and horses have to lead quietly, either by a helper walking at their heads or by a rider from the back of another horse. They also must not mind side helpers walking on each side of them during the lessons.

Finally, the horses must learn to accept or ignore accidental kicks in the sides when being mounted, occasional jerks on the reins, unusual movements by their riders during the exercise periods, having beanbags or balls tossed from their backs during a game, and mounting and dismounting from either side.

Some horses learn faster than others, but by repeating the lessons over and over again with plenty of patience, they can all be taught eventually what is expected of them.

Once a week the horses and ponies in therapeutic programs are ridden by other riders so they can relax and enjoy themselves. It is an easy but boring life for the horses, so weekly trail rides are a vacation for them. A pony that wouldn't think of shaking up a disabled rider will occasionally let off steam with a buck or two during these rides. Horses often pick up bad habits when used only by beginning riders, so this time is also spent schooling them to remind them of their manners.

The leaders and side helpers are often teen-agers from Girl or Boy Scout Troops or 4-H Clubs, who are working toward a community service badge or award. Sometimes they are young people from a church group or from a local high school who are interested in a career working with mentally retarded or physically disabled persons. The time spent as a volunteer gives them a chance to see for themselves what it is like working in this field.

Volunteers should be at least fourteen years old, familiar with horses, dependable, and on time for the classes in which they are to be helping.

Special equipment is required for many of the riding students so that they can learn how to ride safely. In addition to the usual saddle and bridle, there are handholds, safety helmets, safety stirrups, ladder reins, body harnesses, and overhead checks.

English saddles and bridles are used because the English-style rider has more freedom of movement. The horn and high pommel (front) and the cantle (back) of a Western saddle do not allow the rider enough room for exercises, the jumping position, or unusual mounting methods.

Handholds are attached to the front of the saddle to help the beginning student balance himself. Without them, the student would hang on to the reins, jerking and pulling on the pony's mouth.

A safety helmet or hard riding hat is necessary to protect the student's head if he or she falls off. Safety stirrups are also for the student's protection in case of a fall. One side of the iron is made like a regular stirrup, but the outside has a rolled rubber piece attached to the top and bottom of the stirrup. If the student falls, his foot will not get caught in the stirrup because the rubber piece will unsnap with pressure and release the foot. There is almost no chance of being dragged by a spooked or runaway horse when a safety stirrup is used.

Ladder reins are for those who can use only one hand or have trouble holding the reins. Connecting the two reins, like a water-skier's towrope handhold, are three or four rolled-leather or wooden bars. The student holds on to whichever bar is set at the correct length for the gait and pony. The bars make it easier to steer and easier to shorten and lengthen the reins.

The body harness is a four-inch-wide web belt that fastens around the student's waist. A leather handhold is centered in the back for the side helpers to hold on to in case they have to help the student keep his balance in the saddle.

Overhead checks are put on ponies who reach down to nibble at the grass during a lesson. An insecurely seated student might be pulled over his mount's head if the reins were suddenly jerked forward. To prevent this, two pieces of cord are tied to the rings on each side of the front of the saddle, slipped through a ring at the top of the bridle, and snapped to the rings of the bit.

Many students in the Riding for the Handicapped programs also learn how to care for their mounts. They get a chance to brush them and are taught how to put on and take off the saddle and the bridle. Often this is just as much fun as riding, because the children enjoy working around their favorite ponies.

Some of the students learn to ride so well that they are able to ride in activities other than their weekly lessons. Special classes in horse shows, such as those at Thorncroft Equestrian Center in Malvern, Pennsylvania, allow special riders to enter walk and walk-trot competitions. The riders are judged on their ability to control their mounts and on how well horse and rider work together.

In 1976, the Michigan Special Olympics included competitive horseback riding in their games for the first time. Twelve mentally retarded riders from all over Michigan took part and won medals.

Teen-ager Debbie Phillips, of California, is probably the most outstanding disabled horseback rider in our country and even in the world. She was born with her right leg half the length of her left leg. However, after years of hard work and constant practicing, Debbie was able to compete against and win over some of the world's top riders in jumping events in several horse shows, including the International Horse Show in Washington, D.C.

Therapeutic riding stables are nonprofit organizations that continue their work through donations. They also hold fund-raising events to help pay expenses—which include veterinarian and blacksmith fees, feed and bedding for the horses, purchase and care of riding and grooming equipment, and salaries for the riding instructors and stable help.

Sebastian Riding Associates, Inc., puts on an annual Ride-A-Thon, in which riders ask people to contribute money for every mile they ride. Karen, one of the students, rode Pixie, one of the program's ponies, ten miles in a recent Ride-A-Thon. They followed quite a difficult route that included going up and down hills and through streams.

Therapeutic riding programs work closely with doctors and physical therapists. The doctor gives the therapist the child's medical history and the therapist tells the riding instructor what exercises will help the individual student overcome or lessen his problem.

In addition to contributing to the health of students, therapeutic riding is designed to be fun. Games, such as Follow the Leader and Simon Says, are always part of every lesson. The horses and ponies play a big part in them, and the children love their four-legged friends.

After the day's lesson it isn't unusual to see the children give "their special friends" a big hug or slip a carrot into their feedboxes. One little girl at the Sebastian Riding Associates' stable carefully placed a big plastic ring over one of her pony's ears, and said with a wide smile, "Now she's really my angel."

3
Bottlenosed Aquanauts

The United States Navy jet flew steadily on its easterly course. Far below, the whitecaps of Atlantic Ocean waves rolled across the water.

Inside the plane were two bottlenosed aquanauts—two dolphins named Bubbles and Rudy. They lay patiently in slings that hung inside topless wooden crates. The slings, hammock-like stretchers, had openings cut out for the dolphins' flippers.

Hoses controlled by automatic timers periodically sprayed the animals with a fine stream of water to keep their skins from drying out. The dolphins were turned occasionally so their weight was not always pressing on the same side. Bubbles and Rudy didn't really mind being out of water. In fact, they were quite comfortable in their slings.

Bubbles and Rudy were two young, male Atlantic bottlenosed dolphins. They were being sent to find a live bomb that had fallen off a plane into the Atlantic Ocean off the coast of Spain. The dolphins' job was to locate the bomb so it could be recovered.

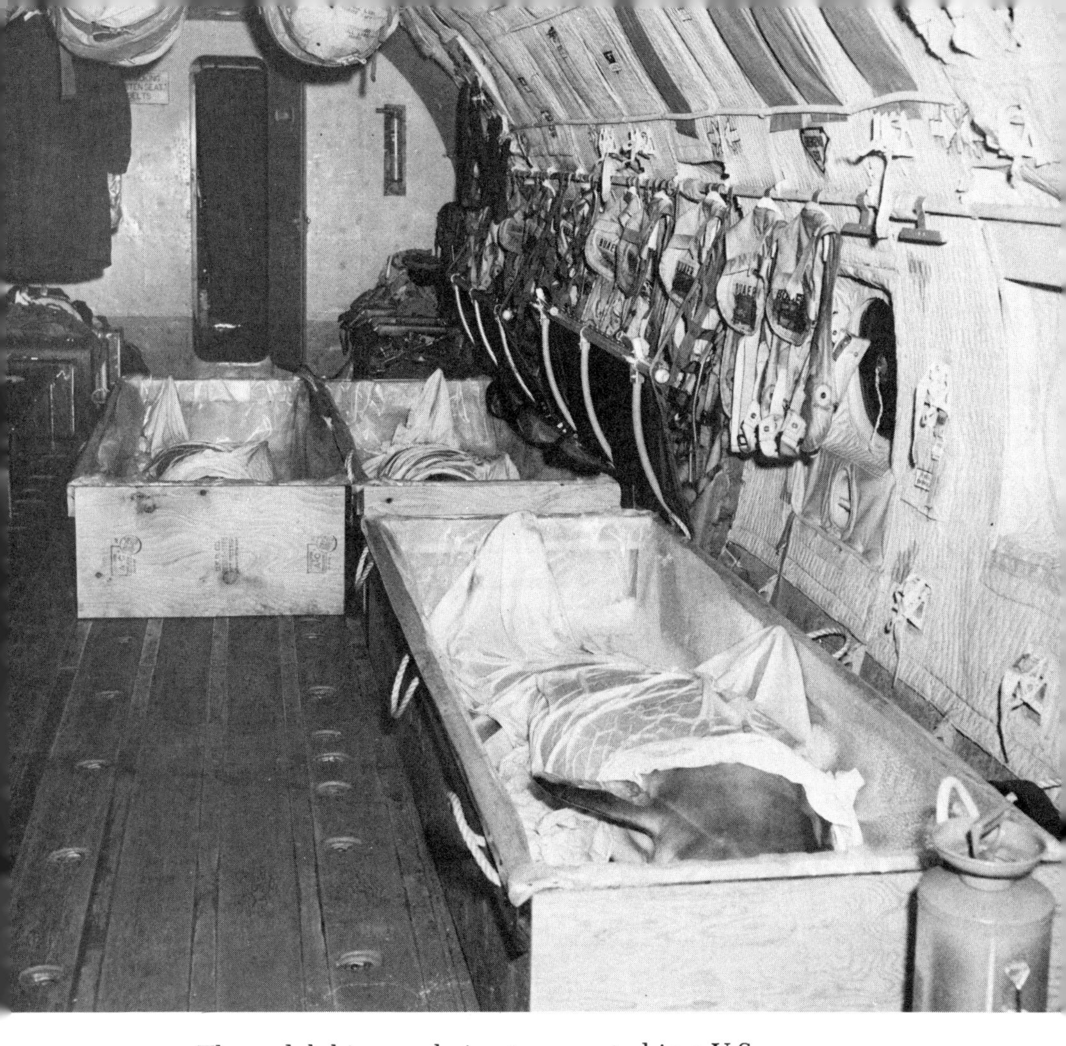

Three dolphins are being transported in a U.S. Navy plane. The sprinkler can to keep their skins wet is in the foreground

After several hours the plane finally touched down at an airport not far from the Spanish coast. Bubbles and Rudy were carried off the plane, and, still in their slings, put into the back of a truck. The drive to the dock, where a boat was waiting for them, wasn't as smooth as the plane ride, but the dolphins just rocked gently in their slings. Dick Hammond, their handler and trainer, sat beside them and sprayed water by hand onto their backs and heads.

"Easy, boys, we'll soon be there," he said soothingly. "Then you'll have some work to do."

Once loaded on the boat, Bubbles and Rudy became a little restless, clicking and whistling to each other. They knew that the ocean was near and their journey would soon be over.

Suddenly the roar of the boat's motors died down to a low rumble and, finally, a purr. Then they stopped. The boat rocked back and forth as the waves passed under and around it.

"O.K., let's get them into the water," Dick called out to the other men.

After attaching the sling that held Bubbles to the boat's crane, the man at the controls carefully hoisted the dolphin up over the side and into the water. Dick was there to help Bubbles wiggle free. Next came Rudy.

"All right, Rudy, just a minute and you'll be on your own, too," Dick said.

As soon as they were loose, the two dolphins skimmed through the waves, leaping high into the air and flopping down to splash Dick as he climbed into a small raft that was tied next to the boat. Their gray, streamlined bodies crisscrossed swiftly underwater as they played with each other.

"O.K., boys, warm up with these," Dick said. He

tossed a couple of fish to each dolphin. The animals leaped up and each caught a fish in midair before snatching the ones that had fallen into the water.

After giving them a few more minutes to play, Dick whistled to the two dolphins. Obediently, they swam alongside the raft so Dick could put on their harnesses. A magnetic ring placed over each dolphin's beak, or nose, would stick to metal objects. The dolphins had been trained to bring these rings up to the bomb. With a flip of their heads they could toss them off their beaks so the magnetic force would hold them in place on the explosive. Then a small float attached to a line would pop up to the surface. Divers could see the line and follow it back down to the bomb. Grappling equipment could then be attached to bring the bomb up to the surface.

Using hand signals, Dick commanded the dolphins to start the search. Both dolphins hyperventilated, taking several deep breaths to draw more oxygen into their bodies, before diving. Then down they plunged into the ocean's blackness.

Once on the bottom, the dolphins circled and weaved in different directions, searching for the bomb. Making a clicking sound, they moved their heads from side to side so they could pick up any clicks that echoed back from their target. Dolphins use echolocation, a kind of sonar, to find objects. Their eyes are almost useless in the underwater darkness. They send high-frequency sound waves through the water and "read" the vibrations that bounce back to them from various objects.

After more than three minutes, Rudy swam up to the surface. Bubbles followed thirty seconds later. They cruised slowly around the boat, waiting for Dick's next command. Dick threw them each a fish, which they gobbled up eagerly.

"O.K., fellows, let's try it again."

At a wave of Dick's hand, the dolphins again hyperventilated and dove to the ocean bottom. This time, both dolphins appeared between the waves before three minutes had passed. The rings were missing from their beaks, and, as the dolphins whistled and squealed for their reward, two little floats popped up on the crest of a wave about twenty-five yards off to the right of the boat.

"Good boys," said Dick, praising them, as he threw a bucket of fish into the water.

Bubbles and Rudy darted back and forth, gulping down the fish as fast as they could. After circling and leaping, both together and separately for several minutes, the dolphins slowed down and swam calmly side by side around the boat. Their backs slid along just below the surface of the water, while their dorsal fins cut through the waves. Occasionally, one would stop, poke its head out of the water, and whistle to Dick.

"All right, let's go home," Dick said as he signaled to Bubbles to swim into the sling he had ready for him. The human divers had already started their descent to the ocean's bottom. In just minutes, Bubbles and Rudy had saved them hours of work searching for the bomb.

* * *

The United States Navy's Marine Mammal Program is training dolphins and other marine mammals to do many jobs underwater. Dolphins have been trained to find lost objects, to rescue lost divers, and to carry tools and messages from the surface to men stationed in underwater labs.

Man has many problems trying to work underwater. He cannot see or hear well, moves slowly, and has to wear bulky, hard-to-handle equipment when going very deep. However, the dolphin can hold his breath for several minutes—as long as six or seven minutes, can swim up to twenty miles an hour and dive to great depths over and over again. Dolphins usually hyperventilate on deep dives, but don't need to for shallow dives. Sonar replaces their eyes, and their great sense of hearing is excellent even underwater.

The body control of a dolphin is amazing. He can leap sixteen or more feet out of the water, turn sharply in the water, come to a dead stop, and coordinate his swimming and leaping perfectly with other dolphins.

The United States Navy program has done most of its work with the Atlantic bottlenosed dolphin, but Pacific bottlenosed dolphins, Pacific white-sided dolphins, Pacific pilot whales, killer whales, Dall's porpoises, and sea lions—commonly called seals—have been used, too.

Tuffy, a 400-pound Atlantic bottlenosed dolphin, is the most famous of the Navy dolphins. He proved that a dolphin could be trained to do useful and important jobs for the man-under-the-sea programs. Tuffy carried small tools and message capsules back and forth between the men on the surface and the aquanauts in Sealab II. In case of an emergency, he could get medical supplies to the aquanauts faster than any underwater elevator.

Tuffy also learned to locate the expensive launching cradles of the Regulus missile. After take-off they would fall back to earth, into the ocean, where they were very difficult to find. It took human divers hours to locate them, but it took Tuffy only seconds. He was able to swim down to 990 feet below the surface—his deepest dive—in three and three fourths minutes.

Buzz Buzz, so named because his signal "to come" was two buzzes, was the first dolphin to be released in the open ocean. Earlier dolphin tests and training had been done in pools, floating pens, and netted-off sections of lagoons or bays. In the open ocean Buzz Buzz went right to work and never strayed or went off "fishing." She proved that trained dolphins could be depended on to do their jobs even when there was no fence to hold them.

In fact, most dolphins who do escape from their pens prefer to stay with their trainer and are easily recaptured. A few dolphins stray, swim back to the beach, or catch and eat fish instead of working, but these are the exceptions. Most continue working until their job is finished.

Buzz, so named because his signal "to come" was one buzz, was trained to follow his handler's boat by the sound of the motor. He also proved how businesslike a dolphin can be. One day while he was following the boat a school of fish swam right by him. He quickly turned, caught one, brought it up to the boat, and tossed it in for his handler to give to him. He wouldn't eat it without permission.

The Atlantic bottlenosed dolphin weighs about 30 pounds and is about three and a half feet long at birth. The average dolphin seen on display is between seven and eight feet long and weighs 300 to 400 pounds. Some dolphins found in our western Atlantic Ocean measure up to ten feet long and weigh 850 pounds. Others in the waters closer to Europe are even bigger.

The life expectancy of a dolphin in captivity is 20 to 25 years. Vicki has set the record so far by living to be over 30 years of age at the Brookfield Zoo in Illinois. She died in 1978.

Atlantic bottlenosed dolphins are gray in color, with a pinkish underside. Their delicate skin is very smooth to touch, and dolphins like to be petted.

Dolphins have a built-in smile that makes them appear friendly and playful. Their personality usually matches their looks. They are friendly and playful—except when angry. Angry dolphins open and shut their mouths as a warning, even though biting is not their usual way of attacking. They will also bob up and down rapidly and swing their heads from side to side to show their annoyance.

Sharks, one of the reasons for anger, will attack a dolphin, especially if he is alone. Most people believe that dolphins are able to outfight sharks, but this is not true. Many captured adult dolphins have nasty scars on their backs from shark bites. Dolphins will attack an enemy in groups, charging with incredible force and ramming their snouts into a shark's body. One "punch" to a shark's liver and it has usually had enough.

Most trained dolphins were once wild. They must be captured from the oceans, because breeding dolphins in captivity has not been very successful. To catch dolphins, boats go out and chase them into big nets. Then the nets are hauled in and the dolphins are pulled into the boat. Once entangled in the net, the dolphins must be pulled out of the water and up into the boat immediately or they will drown. It takes great strength and skill to haul a struggling 300 to 400 pound dolphin aboard a boat within a few minutes. Only the dolphins that are two to three years old and healthy are kept. Those that are sick, too young, too old, or injured are released.

The next test the captured dolphin must pass is how quickly he becomes tame. If he begins eating right away in captivity, he is a possible candidate for the Marine Mammal Program. If he refuses to eat, he is put back in the ocean so he will not starve to death.

Dolphins are afraid of people at first, so the trainers will stroke and rub the frightened animals to calm them. The trainers will also get right into the water with the newly captured dolphins to force them to make physical contact with man. It doesn't take long for a dolphin who adjusts to captivity to start eating from the trainer's hand.

Just like people, some dolphins are smarter and

A recently captured dolphin is lowered into his new home in preparation for training in the U.S. Navy's Marine Mammal Program

A dolphin will lift himself up out of the water to look for his trainer's next command

easier to teach than others. Those that are quick to learn will go on to a more advanced training program than those that are easily distracted, slow to learn, or stubborn and too playful.

Bottlenosed dolphins are fairly easy to keep in good health if they are given proper attention. The biggest problem with their care is keeping the seawater clean and in good condition. An adequate supply of quality fish is also necessary. A dolphin will eat 15 to 25 pounds daily of mackerel, blue runner, and other ocean fish. Dolphins receive vitamins, too, to help keep them from getting sick.

The water in their tanks doesn't have to be real seawater and doesn't even have to be very salty, but the temperature should be between 55° F and 85° F. In 1963, Illinois' Brookfield Zoo, Philadelphia's Aquarama, and Animal Behavior Enterprises, Inc., in Hot Springs, Arkansas, were the first to bring dolphins inland and keep them in artificial saltwater. Today, there are many dolphins living hundreds of miles away from the oceans.

Dolphins usually enjoy being trained to do tricks and perform jobs. They like challenges and games, and will even invent their own when bored. In the wild, a dolphin will "race" a boat or body-surf on large waves that roll in to shore. A ball or just a feather that blows into the tank will be tossed and carried about for hours by a playful dolphin in captivity. However, a small object, such as a coin or a watch, accidentally dropped into the tank, may be swallowed and cause a serious problem, even death, to a dolphin.

A properly trained dolphin will respond quickly to the signals of his handler. Training a dolphin requires common sense, patience, and using proper signals regularly. The animal must not be frightened or confused. A dolphin is never punished if he does something wrong, but he is always rewarded with a pat or a treat —a special fish that he likes best of all, whenever he does what he is told to do.

It is very difficult, if not impossible, to make a large

male dolphin do anything that he doesn't want to do. The trick is to get him to want to do it. If a dolphin refuses to do what he is told, the trainer may stop the lesson and walk away. This will usually cause the dolphin to swim to the side of his tank, stick his head out of the water, and "call" for his trainer to come back. Dolphins like human companionship and some even prefer people to fellow dolphins. They are able to recognize the people who work with them and know their voices, too.

All dolphins have their own personalities. Some will work with only one trainer, others will obey any trainer. Some dolphins get very jealous if their trainer pays too much attention to another dolphin, and others work well with a partner. If a dolphin is annoyed, he may slap the water with his tail, drenching his trainer, or give him a butt or even a bite. Tuffy did not feel he had been rewarded fast enough for a good job one day, so he immediately swam to the surface and bit his regular trainer.

Dolphins may be trained to work by sonic signals, buzzers, gongs, bells, hand signals, or voice commands. They learn tricks from other dolphins, too. A good performer is a dolphin who has some natural spirit and who shows a willingness to be handled. Once a dolphin learns a trick he will remember it for years. One dolphin did one of his acts immediately on command although nearly ten years had passed since he had last performed it.

Dolphins enjoy being petted

Sea lions have also been trained to do work for man. They seldom develop the close friendships with humans that dolphins do, but they pick up training very quickly because of their great liking for food rewards. Project Quick Find uses sea lions to recover equipment from the bottom of the ocean.

When sea lions make dives in the open sea, they have to wear muzzles to keep them from trying to catch and eat fish. One sea lion dove down to a depth of 750 feet during a practice session.

Fatman, a four-year-old sea lion looking for dummy

antisubmarine rockets off the Norfolk, Virginia, coast, recently went AWOL (Absent Without Leave). The wanderer left the Navy's Project Quick Find Sea Lion Recovery and turned up three hundred miles away off the New Jersey coast. Easily recaptured, he was returned to his trainer in Norfolk, where he seemed quite happy to be with people and have regular meals again.

Pilot and killer whales are also used for recoveries of dummy or fake torpedoes and other objects. Ahab, a 17-foot killer whale, was trained to put a clamp on a dummy torpedo so it could be brought up to the surface. His deepest dive was to a depth of 850 feet, where he put a clamp on a torpedo in a little over seven and one half minutes.

Ahab would also tow a swimmer and allow handling and brushing. His trainer could step on his back from the side of the tank and walk around, scrubbing him with a long-handled brush, while Ahab slowly turned over.

Morgan, a 13-foot pilot whale, who weighed about 1,200 pounds, dove to a depth of 1,654 feet in a little over twelve and a half minutes.

Stories about dolphins and how they have made friends with and even rescued drowning people have continually been told and retold. A popular movie and TV series, *Flipper,* was based on these almost unbelievable tales.

During World War II, it was reported that a dolphin pushed a raft carrying six American airmen shot down

A dolphin can leap more than sixteen feet out of water

by the Japanese to a small island in the Pacific Ocean.

In 1955, Opo, a friendly dolphin, came daily to a beach in New Zealand to play with the children. Tag, fetch-the-ball, and ring-around-the-dolphin were some of the games she played with them. Opo also gave Joy, her thirteen-year-old special friend, rides on her back. Although the government passed a law to protect Opo, she was accidentally killed by three boys setting off dynamite underwater to catch fish.

Another dolphin, Donald, has been swimming up and down Great Britain's western coasts, making friends with people, for about five years. He has twice saved men from drowning by holding them up on the surface until help arrived. He gives rides, takes rides himself by hanging on to an aquaplane—a board pulled along by a motorboat—with his mouth, points out fish to fishermen, and finds lost objects. Although he is twelve feet long, and weighs 880 pounds, Donald seems only interested in having fun. The schoolchildren of Cornwall, on England's southwestern tip, keep an eye on Donald through Project Dolphin Watch.

In the future, dolphins might be used to find oil leaks in the oil industry's underwater pipes, guard industrial areas against sabotage, prevent smuggling by sounding an alarm if a strange boat tries to dock in an off-limits area; gather information, with cameras and other equipment, about the ocean's bottom for scientists; lead rescuers to floating survivors of a ship or plane accident; and warn swimmers of approaching sharks.

According to Jacques-Yves Cousteau, two dolphins, Dimple and Haig, have already been taught to chase sharks away from swimmers' beaches in South Africa.

With all these possibilities, the dolphin could easily become the second most useful of all our animal friends. Only the dog would then perform more jobs for his human masters than the dolphin.

4
Animals in Advertisements

"Lights, action, roll 'em."

The shaggy little dog lay quietly on the braided rug in front of the television. Curled up in a ball, he rested his nose on his hind feet and the tip of his brown-and-white tail.

Suddenly, Scruffy lifted his head, pricked up his ears and looked toward the doorway. Some strange noises were coming from the front hall. Before he could get up to see what was going on, a miniature mule team pulling a tiny red-and-white-checkered chuck wagon charged around the corner.

The little dog sat right up with as surprised a look on his face as any human actor could make. He watched for a moment as the team and wagon darted under the edge of the living room rug. Finally, he couldn't stand it any longer.

Barking excitedly, Scruffy gave chase. The mules popped out from under the other side of the rug and galloped into the dining room. The dog dashed after them. Under the table and between the chairs the

Scruffy watches the Chuck Wagon race by him and duck under the living room rug during a television advertisement

mules, swaying wagon, and dog went, twisting and turning around the legs of the furniture.

The driver of the wagon cracked his long, black whip and shouted encouragement to his team.

"Hi yup there, mules, get along."

The chase roared into the kitchen. The wagon's wheels rumbled across the floor as the little dog appeared to be catching up. One more step and Scruffy would be able to grab them. Suddenly, the mules leaped right through the closed door of a kitchen cabinet and disappeared, pulling the chuck wagon in after them.

The little dog skidded to a stop and stood in front of the cabinet with his head cocked and a puzzled look on his face. After waiting for a few seconds to see if the mules and wagon would reappear, he began barking angrily.

The dog's owner walked into the kitchen and over to the cabinet, where he took out a box of Chuck Wagon dog food. Tail wagging happily, Scruffy quickly sat up on his hind legs and waved his front paws in the air, begging for his dinner.

"Cut!"

* * *

Scruffy, the dog who stars in the Chuck Wagon ads on television, is a veteran of commercials. He has appeared in many different advertisements, although he is recognized by most people as the dog who chases the chuck wagon across television screens.

Other well-known animals that have appeared with various products, either on television or in magazine and newspaper ads, are the Budweiser horses, the Snak Pak palomino, Morris the 9-Lives cat, the white stallion for White Horse Scotch whisky, the cougar for Lincoln-Mercury, the kittens for Little Friskies cat food, Smokey the bear for Fire Prevention Week, and Jackie, the lion that roars at the beginning of all MGM (Metro-Goldwyn-Mayer) movies.

Animals are natural scene stealers and lure viewers' eyes to themselves without even trying. By drawing the attention of the public to themselves they also draw attention to the item they are selling, whether it is pet food or cars. When a monstrous bull charges a man sitting at a café table enjoying a can of beer, viewers can't help watching to see what happens. They also notice the name of the product.

When a huge gorilla throws a suitcase around his cage and jumps up and down on it, most people watch this commercial, too. Fascinated by the ape and his actions, they can't help seeing the brand name of the product being advertised.

Perhaps the oldest and most successful animal advertisers are the eight-horse teams of massive Clydesdales that advertise beer and travel thousands of miles across the country every year. The Clydesdale show was started in 1933 by August Busch, Jr. The teams pull their 3½-ton, red-and-white wagons in parades, at horse shows, around shopping center parking lots, at rodeos, along racetracks, and through fairgrounds. Moving at a fast, high-stepping trot in perfect time with each other, they twist and turn in circles, diagonals, and figure eights.

The Budweiser 8-horse hitch shows agility while making a small circle

These kings of the horse world are six or more feet high at the top of their backs and weigh over two thousand pounds each. There are two teams, called hitches. The Show Hitch, based in St. Louis, Missouri, covers the western part of the country. The Merrimack Hitch, based in Merrimack, New Hampshire, goes to all the events in the eastern United States.

The horses range in age from three to eleven years. Retired hitch horses can be seen at Anheuser-Busch parks, such as the one outside Williamsburg, Virginia.

The Clydesdales, a breed of draft horses originally from Scotland, start their training when they are about two years old. In addition to learning how to work together in harness, they must also stand quietly and ignore the noise and confusion around them. They must not mind people trying to pat their noses or shoulders, and they should not get upset if balloons pop, firecrackers bang, or horns blast practically in their ears.

To help protect their legs and feet from the long hours of standing or trotting on hard roads, leather pads have been put between the four-pound iron shoes and their hoofs.

The hitches travel in two caravans—each with three white vans that carry the horses, their portable stalls, the harness, wagon, equipment, and personal belongings of their handlers. Ten horses make the trip, so that the two extras can be used as substitutes if one of the regulars gets sick or needs a rest.

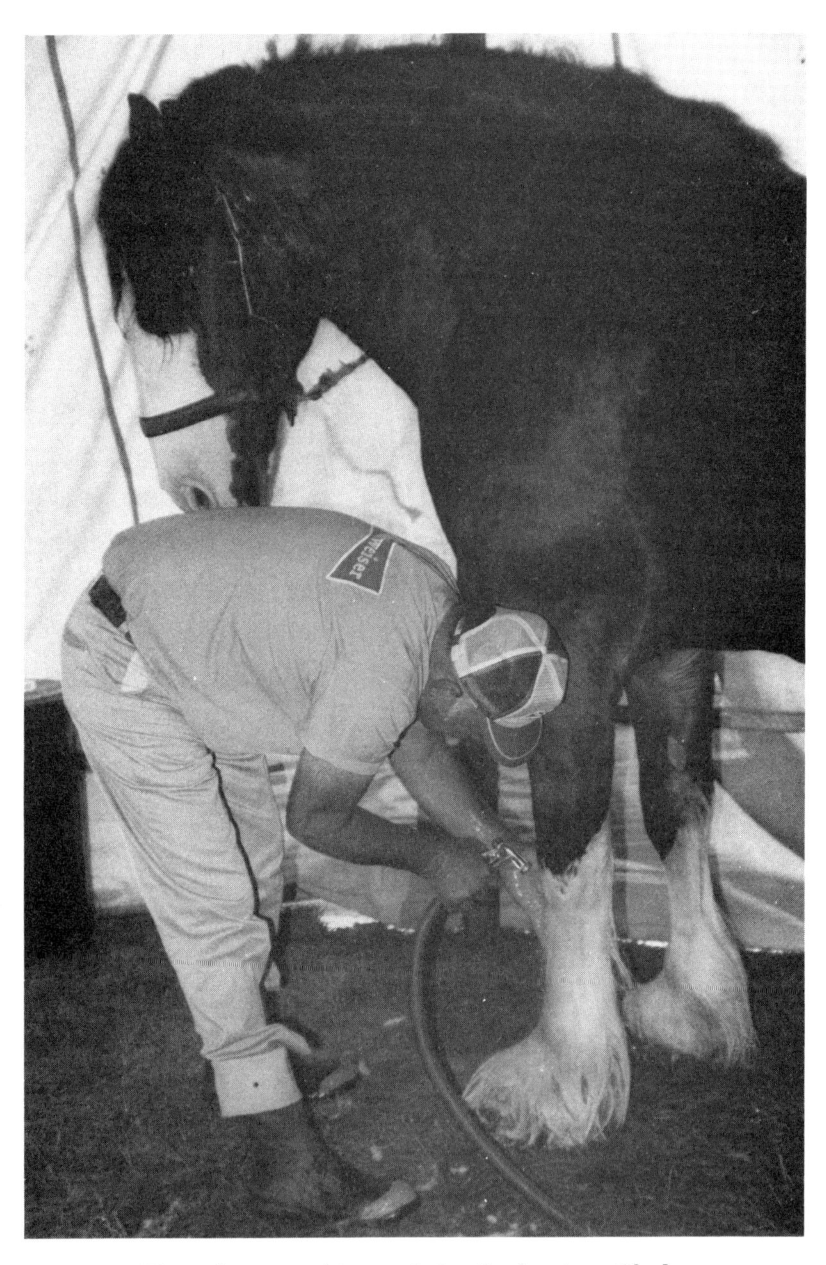

The white stockings of the Budweiser Clydesdales are washed before every performance

Preparing the Clydesdales for a performance takes five hours of steady work by six men. White stockings are washed, coats are brushed and vacuumed, and manes and tails are braided with red and white ribbons.

It takes the handlers thirty minutes to hitch the horses up to the wagon. Their black leather-and-brass harness has been custom-made for the team. Each horse's set weighs over 130 pounds. Two men work constantly every day cleaning, oiling, and polishing every piece of it.

Another well-known animal advertiser is Morris, star of 9-Lives cat food commercials. Known for his finicky behavior and lack of interest in anything but his favorite cat food, Morris became the personality cat to TV viewers.

The original Morris was found in an animal shelter in the Chicago area in 1966 by animal trainer Bob Martwick. Not so pretty as other cats, Morris was a battle-scarred, fifteen-pound, orange tomcat, more used to meals from a trash can than a dish. However, his self-assurance and super-cool attitude at the auditions earned him the 9-Lives role.

In addition to making commercials, Morris starred in the movie *Shamus* with Burt Reynolds. In 1973, he was awarded the Patsy, which is equal to the Oscar given to human actors.

During his most popular years, fans were able to buy Morris bowls, Morris T-shirts, Morris calendars, and even the story of his life, *Morris: An Intimate Biogra-*

Morris the 9-Lives cat looks over some maps before starting a promotional tour

phy. The good life was not always a bowl of tuna, however, and although Morris lived a life of ease on his six-and-a-half-acre suburban Chicago estate, it was reported that he had round-the-clock guards to protect him from "catnappers."

The original Morris's nine lives ended in July 1978, when he died of heart failure at the age of seventeen. An understudy has been selected to carry on the role of Morris in 9-Lives commercials.

Chauncey, the original Lincoln-Mercury cougar, starred in the Sign-of-the-Cat commercials. He was raised from a cub by animal trainers Pat and Ted Derby on their 300-acre farm for wild animal orphans in California.

Chauncey learned to snarl when Mr. Derby gave him the signal. He knew that whenever he saw the Lincoln-Mercury sign he was to leap up on top of it and let loose with his famous snarling scream. He worked by voice

Chauncey, the Lincoln-Mercury cougar, looks toward owner Ted Derby for the command to "sit" on the roof of a Mercury Cougar

Larro
Feeds

OPEN YOUR
SAVINGS
ACCOUNT
AT COAST
9TH & HILL

This baseball-playing chicken helps to advertise General Mills' Larro Feeds. Buck Bunny was the star of the record-setting commercial for Coast Federal Savings in Los Angeles, California

directions and knew he had done something wrong when told "No."

As he got older and heavier (200 pounds), Chauncey was usually used only in the scenes where the cougar was supposed to sit or lie down. Harold and Herman took over for Chauncey when the scenes required physical activity, such as jumping and running. They were twin cougars and younger than Chauncey.

The Derbys always took three cougars, all about the same size, on every filming assignment. This way all the cats had plenty of time during the day to rest in their air-conditioned cages.

Although the Derbys stayed out of camera range, they both stayed near the cougars during filming. When Chauncey was in a car with a model, Mrs. Derby always lay on the floor of the car so she was right there if there was a problem. If the big cat had to be placed in another position, she did it. The model never moved him.

As part of the cougars' training, the Derbys let them

ride in their station wagon up and down California freeways to get them used to noises and people. Wild cougars are not really trained, according to Mr. Derby. They are only made comfortable and kept as happy as possible so that they learn to trust people, fear nothing, and will do as they are asked.

Chauncey and his fellow cougars at the Derbys' animal farm live in electronically heated cages. The average daily meal for an adult cougar is seven pounds of horsemeat and chicken necks, sprinkled with vitamins.

Chauncey died and was replaced by Christopher in the commercials. However, the younger cougar did not have the patience or the desire to please that Chauncey had when performing. Christopher appeared in the Sign-of-the-Cat commercials for Lincoln-Mercury, too, and he was in an episode of the TV series *The Six Million Dollar Man,* but eventually the tight shooting schedules angered him. Christopher rebelled and was retired from show business.

One of the most lovable animals in advertising is Scruffy, the little dog described at the beginning of this chapter. He is well known for his acting in the Chuck Wagon commercials and seems to enjoy chasing the mules and wagon.

Scruffy was found by animal trainer Carl Miller in the Sherman Way Animal Shelter in North Hollywood, California. Mr. Miller felt sorry for him, so he paid $7.50 and took him home. Scruffy is a happy-go-lucky dog, described by Mr. Miller as "an accident looking for something to happen." He was full of mischief when first brought home. He ran away when being brushed, dug holes in the yard, and pulled clothes off the clothesline.

Scruffy was also naturally "toy crazy," which helped in training him. His reward for doing a trick correctly was to be able to play with a squeak toy.

In *The Ghost and Mrs. Muir,* Scruffy was supposed to walk alongside a boy on a beach. Scruffy didn't know the boy and wouldn't have walked with him, so they put his little squeak mouse in the boy's pocket. The boy would press his arm against the toy to make it squeak as he walked. Scruffy stayed right at his side, ears cocked and eyes on the boy's pocket.

For the Chuck Wagon commercials, Mr. Miller would let Scruffy see him put a squeak toy in the kitchen cabinet. Then he would bring him into another room and let him go. The cameraman would film the little dog racing around the corner, dashing across the kitchen, and skidding to a stop in front of the cabinet. Scruffy would then sit watching the door until Mr. Miller opened it so he could get his toy. The mule team and wagon were filmed at another time and place and spliced into the first film so it looked as if Scruffy were chasing them.

Scruffy has been working since 1968. In addition to appearing in various commercials, he has been on *Love —American Style, Cannon, Barnaby Jones, The Red Hand Gang,* and several Movies of the Week. His tricks are almost unlimited.

One of the hardest things for Scruffy to do is to go through the same scene over and over again. Directors rarely feel that a scene is shot perfectly on the first take, so they may order several retakes. The dog knows he has done it right the first time and can't understand why he has to do it again . . . and again. It is very boring and tiring for him, so Mr. Miller tries to make the filming as interesting as possible. He will let Scruffy do another trick, play with him, give him some exercise, or allow him to rest in his "dressing room" in between takes.

A dog likes affection and will work for praise alone from his master. Cats and most other animals work for food. They learn to do a trick when a buzzer, clicker, or bell signals them to do it. They are rewarded with a piece of food for doing the trick correctly. Since they won't work when they are not hungry, they are fed only half of their daily food portion at mealtimes. The rest is given to them as rewards during training or working sessions.

Doubles are used for animals that have to gobble up a bowl of pet food for a commercial. Once a dog has eaten one bowl he will not do it again until hungry. This is when the double steps in and takes his place for reshooting.

A lot of time is spent teaching a cat her first trick, but after she gets the idea, she will learn new tricks even faster than a dog. Animal trainer Frank Inn says you can train a dog, but you have to "con" or fool a cat.

For a Little Friskies cat food commercial, animal trainer Moe DiSesso taught four cats to play a grand piano at the Hollywood Bowl. They learned to sit on a piano bench with their front paws on the keys. When Mr. DiSesso said, "Play, play, play," they would bang on the keys as if playing a song.

Mr. DiSesso started with fifteen cats, but dropped five after three days because they did not pay attention to him. In three weeks he trained the remaining ten cats to play the piano, then picked the four best for the commercial.

Mr. DiSesso also trained a cat to open her own can of cat food with an electric can opener. However, cat food companies did not want to use this trick because they felt it was so good that people would think it was a fake. Finally, a cat food company in Canada bought the stunt.

Although animal advertisers may never use the product they are advertising, they are very good at making people aware of it. The next time you are watching television, count the number of commercials that have animals in them. The animals may not be stars like Morris or Scruffy, but they do stand out even if they are only in the background. Look for a bucking bronco, a pony pulling a cart, several different breeds of dogs racing for a dish of their favorite dog food, a cat lying in the lap of a movie star who is explaining a product, an actress cantering her horse along a beach, or a dog standing next to a model who is demonstrating the latest fashions.

They are all attention getters—proven eye catchers and very valuable in the fields of advertising and sales.

5
Ears for the Deaf

It was a bright, clear moonlit night. Only the soft glow of streetlights showed that the town had not been deserted.

Harvest Acres was a typical suburban housing development. However, not all its residents were typical suburbanites. At 117 Dogwood Lane all was quiet. Everyone had gone to bed hours earlier. Suddenly, a wisp of smoke began leaking out of a downstairs window. Barely visible at first, it soon became a never-ending gray snake, twisting and spiraling up into the sky.

Jamie was curled up on his rug beside his master's bed. His black nose was tucked under the feathery tip of his long, black tail. Sighing heavily, he rolled back and stretched out on his side. Suddenly his nostrils started twitching as the slight smell of smoke drifted up into the bedroom. Jamie sat up. Something was wrong!

Just then the smoke alarm went off in the kitchen. Springing to his feet, the small black dog began licking his master's face. Bob pushed him away and mumbled in his sleep. Jamie became frantic. The room was beginning to fill with smoke now. The alarm continued to beep.

Finally, Jamie jumped up on the bed. Whining, he nuzzled and pushed at Bob.

"O.K., O.K., I'm awake," groaned Bob, still half asleep. Then he smelled the smoke, too.

"Oh, my gosh, that's smoke!"

Bob leaped from the bed and rushed for the bedroom door. But smoke was pouring out from under the door. He ran to the bathroom, grabbed a towel, and soaked it with water in the sink. Then he dashed to the door again.

Jamie anxiously ran back and forth from the door to Bob, whining and barking. He knew they had to get out of the house—and fast.

Bob felt the door for heat, then cautiously opened it. The stairway was filled with smoke, but no flames. Bob put the towel over his head and hung on to Jamie's collar.

"O.K., boy, let's go," he said.

Jamie put his head down and started across the hall to the stairway. Keeping one hand on Jamie's collar and his head close to the floor, Bob crawled along as fast as he could next to the dog. Down the stairs they went, bumping into the walls and each other as they struggled to keep going. The smoke was thicker now and neither Jamie nor Bob could see anything.

At the bottom of the stairs they turned toward the front door. Their lungs were aching now and their eyes were streaming with tears from the smoke.

At last, Bob yanked the door open. Together dog and man plunged outside, coughing and choking as they staggered to the sidewalk. Lights were blinking on in houses all up and down the street. The smoke alarm could still be heard beeping in Bob's kitchen. Echoing in the distance was the wail of fire engine sirens. Neighbors rushed out of their houses, buttoning up coats and robes as they gathered around Bob and Jamie.

"What's going on?"

"What happened?"

"Is anybody hurt?"

Everyone had questions, but Bob didn't hear them. In fact, Bob couldn't hear his alarm or the approaching fire engines either. Bob was deaf. He had been without hearing all of his life.

Thankful that his wife, Ann, and their two children were visiting Ann's parents, Bob knelt down and gave Jamie a big hug. He knew that if it hadn't been for the little dog, he probably would not have awakened in time to escape the deadly smoke. It would have filled his lungs and caused his death. His "hearing dog" was truly a lifesaver.

* * *

Jamie's story could have been told about any one of the seventy-five or more other hearing dogs that have been trained to aid their hearing-impaired or deaf owners.

There are over thirteen million people in the United States with hearing problems. Almost two million of them are completely deaf. The sounds that alert others to danger and emergencies, such as a fire alarm or a baby's cry, or various sounds in everyday life, such as an alarm clock or a doorbell, go unheard by these people.

To assist them, the American Humane Association has developed the Hearing Dog Program. Graduates of this program are trained to take the place of their masters' ears. Already, dogs have been placed with hearing-impaired people all over the country from Connecticut to California and from Minnesota to Texas.

The Hearing Dog Program works closely with other organizations formed to help those with hearing disabilities, such as the National Association of the Deaf. It was started by the Minnesota Society for the Prevention of Cruelty to Animals. The Society developed a training program that produced six trained hearing dogs. Publicity in newspaper and magazine articles and on television brought hundreds of requests for these dogs. Unable to handle the demand to expand the program, the Minnesota Society turned it over to the American Humane Association in 1976.

The Hearing Dog Program is managed by the American Humane Association's Director of Animal Protection. The staff is made up of professionals who are experienced with both deaf people and animal training.

The training is done at the Colorado Humane Society in Henderson, Colorado. There is room there to keep twenty-four dogs, and enough space to train them, too. The program is nonprofit and is made possible through donations by individuals and groups.

The dogs trained in the Hearing Dog Program come from local animal shelters. They are animals without homes that probably would have been put to sleep. Now, through this program, they are finding love and a reason for living.

Hearing Dogs can be any breed or a combination of any of the breeds. They can be any size, but should be young enough to provide several years of aid to their masters after being trained. Those accepted in the program are between six and eighteen months old.

The first two months of training are spent learning the basic obedience commands—come, heel, sit, down, stay. The next two months are for sound awareness training—teaching the dog to listen for and react to different sounds.

Sound awareness training depends on the needs of the dog's future mistress or master. Applicants are asked to name the three most important sounds to which they want their dog to alert them. These sounds can be an alarm clock, a doorbell, smoke alarm, baby's cry, door knocks, telephone, intruders, and even the whistle of a teakettle.

Randy, a black mongrel that was part collie and part Labrador retriever, became so sensitive to sounds that he even "told" his master, Gar, when the faucet was dripping or the record player had been left on.

Randy had been brought into the Colorado Humane Society after he was found with a choke collar so tight that it was embedded in the flesh of his neck. The dog had scars from it for the rest of his life and couldn't be trained in the usual way because of this cruel and painful experience.

To teach a dog to wake his sleeping master when the alarm clock rings is one of the more popular requests. The dog is first made to lie beside his "master's" bed. When the alarm goes off, the trainer, who is pretending to sleep on the bed, pulls the dog to him by his leash. Then the trainer pets and praises him. This is repeated over and over again until he gets the idea to come to his master whenever the alarm rings.

A treat is hidden under the "sleeping" trainer's neck to teach the dog to wake his master by nuzzling his neck. When the alarm rings, the dog pushes and pokes with his nose to get the treat and is rewarded with both the treat and more praise.

Teaching a dog to alert his master to the ringing of the phone is also a popular request. A treat is put under the phone and the dog is taught that he may go and get it only when the phone rings. Then he is called to the trainer for praise and petting. He soon learns to go to the phone and then to his master when the ringing starts. A special telephone—one that rings more often than a regular phone—is used during this training.

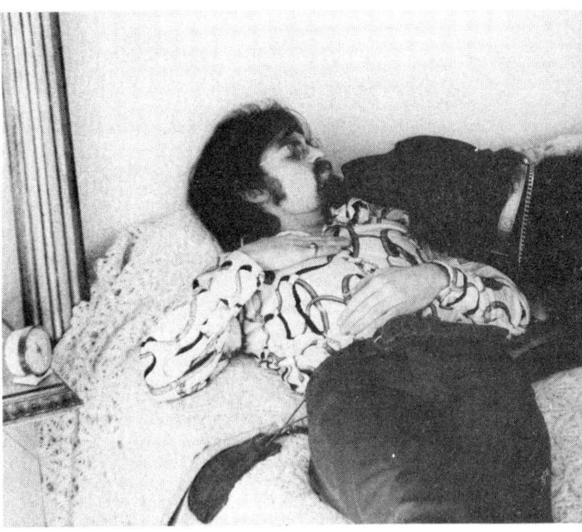

Hearing Dogs can be trained to alert their deaf owners to the whistle of a boiling tea kettle and the ringing of an alarm clock

Training a dog to "tell" his master when the doorbell rings is one of the more difficult jobs. A pulley system is used to pull the dog to the door when the bell rings. Then he is called to the trainer and rewarded. This is repeated until he learns to go back and forth between his master and the door when someone rings the bell or knocks on the door.

To alert their masters to the clanging of fire alarms and the beeping of smoke alarms, the hearing dogs are trained to run back and forth from the alarm to their masters. To get their masters' attention, they will even jump up on them to warn them of the emergency.

Hearing dogs can also be trained to pick up a wallet or keys if they drop out of a pocket without being heard. Those dogs that must alert their mistress to a baby's cry are trained with a tape recording of a baby crying. The recorder is placed next to a doll in a crib and played so the dog can hear it and warn his mistress.

All training is done in short sessions. Several fifteen-minute training sessions are much more satisfactory than one or two hour-long periods. The dog will get bored and begin to dislike training if the sessions are too long.

Every command that is performed correctly is followed by praise and petting. A dog that doesn't do what he is told to do or ignores a command is not beaten or whipped. A stern "no" and a yank on the leash is the only punishment. The sound training is made to be an enjoyable game for the dogs. They soon try to outrun their trainers to the source of the sound and stand there happily wagging their tails, if they arrive ahead of the trainer.

Although it has been estimated that it costs about $2,000 to fully train a hearing dog, these dogs are given without charge to deaf applicants who qualify. Those that complete obedience training, but are not good for auditory training, are placed as companion dogs with other types of disabled people.

People who want to have a hearing dog must apply for one. Older, single people who live alone are the first to get a hearing dog. Adults living with others and young children who can depend on their parents are not usually given a hearing dog.

After an applicant is accepted, he or she must wait from four to twelve months before receiving a dog. A trained hearing dog must become accustomed to its new home, its new master, and the home alarm clock, telephone, doorbell, and other sounds. Not all clocks, phones, and bells sound alike. A trainer accompanies the dog to its new home so he can introduce the dog to its new owner and so he can teach the deaf person how to work with and understand the dog.

One of the more unusual requests came from a deaf and blind couple who needed a hearing dog so they could find each other in their own house. The dog also had to be taught never to touch any food on the table and to eat only what was in his own dish. Since they couldn't see or hear him, the deaf and blind couple wouldn't be able to prevent the dog from eating their dinners, too.

Another request came from a deaf man who did plastering work. He didn't like it when people came up behind him while he was working, because they would startle him and could cause him to fall off his ladder or the scaffolding. He wanted a dog that would warn him of others approaching.

Hearing dogs have to be worked constantly. They have to have their memories refreshed at least once a week or they will forget their training. Just a few minutes of going over the commands and sound awareness training is all that is necessary.

A dog works best when he knows that his master depends on him. If there are other people around who answer the phone or door or alert the deaf person to other sounds, the dog will become lazy and soon stop working altogether.

After a dog has lived with his new master and shows that he is reliable, he is tested by an American Humane Association representative from the Hearing Dog Program. If he proves himself worthy of the title, he is awarded a bright-red collar and leash with "HEARING DOG" written in black letters on them. This tells others that he is a certified hearing dog.

The training center in Colorado is the only one of its kind in the United States. A manual of training techniques is to be put together in the future. This manual will go to regional centers of the Hearing Dog Program. The centers will be set up all over the country in connection with local humane societies.

The American Humane Association hopes that pet companions and pet helpers will also become available for use by the elderly and the mentally ill in the near future.

Flyer, a black, flat-coated retriever, is an example of the way a hearing dog can help its master. Tied to a tree and abandoned by her original owners, Flyer now attends classes with her new master at a county community college. Her master is presently trying to teach her to lift up a paw when his name is called so that he can answer roll call or a teacher's questions. From being an unwanted dog scheduled to be put to sleep, Flyer became a valued helper. She is proof of what one can expect from a canine reject who receives proper training and plenty of love.

6
Monkey Talk

The sun has already been shining down on the little farming community west of Philadelphia for a couple of hours. Farmers with their tractors or teams of mules and horses are busy trying to get most of their work done before the midday heat forces them to stop.

From one of the narrow little roads that splits the plowed countryside, a long driveway curves back to the Primate Center. The gray concrete block building sits in a grove of tall trees with a small, man-made pond in front of it. High concrete block walls curve out and around in a large semicircle from the back of the building to form an outdoor compound.

It is still quiet here. Only the twittering of birds and the occasional harsh squawk of a pheasant break the silence.

Just before nine o'clock a young woman in a white lab coat enters the compound from the back door of the building. Amy carries four pans of food, which she puts down in the grass a couple of feet apart not far from the door. Then she turns and goes back inside.

As soon as the door shuts after her, the building comes alive with noise. Hoots and screams seep out and echo against the walls of the compound.

In a few minutes Amy returns, but this time she isn't alone. Clinging to her hand is a small black chimpanzee.

"Come on, Burt. The others are coming," Amy says to the chimp, who keeps looking back inside for his companions.

Just then three more chimpanzees, Sadie, Luvie, and Jessie, charge through the door, pulling their trainers along with them.

"O.K., kids, dig in," says one of the trainers.

Scrambling to their own pans, the chimps squat down beside them and begin to eat.

Breakfast is one piece of fruit (usually banana or apple) and some monkey chow pellets. The chimps will have this same menu for dinner at about four o'clock in the afternoon. During the day, while being tested, they will be given yogurt, candy, and more fruit.

Burt and Sadie finish eating first and move away from the others. Sitting close together, they begin picking at each other's fur. Luvie looks up at them and decides to make sure their pans are really empty. Leaving her own still half full, she walks over to inspect the other two.

Burt sees his chance and darts over to grab one of her monkey pellets. Then, putting his head down and grunting playfully, he gallops off on all fours, holding the pellet between his teeth. Luvie screams and begins to chase him, but remembers that her pan is again unguarded. She whirls around and dashes back just in time to prevent Sadie from stealing something, too.

Jessie ignores the others. She finishes eating and casually turns her pan over before wandering over to Burt and Sadie. Luvie finally joins them and the four chimps roam around the compound together looking for tasty insects or juicy leaves of a favorite weed.

The compound is both a testing ground and a playground for the chimps. It is a large grassy area with paths crisscrossing each other and bare spots where the chimps relax in the shade. It has a tire swing, big cement pipes to hide in or crawl through, two raised wooden platforms, poles with ropes hanging between them, a huge, weathered tree stump, and a couple of piles of rocks.

Moving along the wall, the chimps continue searching through the weeds, often stopping to roll around and scuffle together or to pick at one another's fur. Sadie finds a flower, puts it between her lips, and passes it to Burt. He grasps it with his lips, then pulls it out of his mouth and examines it. Satisfied, he shoves it back into his mouth, chews, and swallows it.

The chimps rarely just sit. They are always playing—somersaulting, twirling around in circles, swinging a stick, throwing stones, or digging in the dirt. They may stop for a few minutes, but only to think of their next move.

The chimpanzees like to play Hide-and-Seek or King of the Mountain on the pile of pipes in their compound

An old tennis ball that Burt finds causes a few problems at first because everyone wants it. But after quarreling over it for a few minutes, the other chimps lose interest and leave Burt alone with it. He bounces it against the wall a few times, then decides that since no one else wants it, he doesn't either, and he runs off after the others.

The testing begins at ten o'clock. The purpose of the testing is to study communication between two chimpanzees and to learn more about how they "talk" to each other.

Depending on the test, only one or two of the chimps are taken inside at a time to be tested. All the chimps enjoy the testing and rush to the door when it opens, hoping to be the one picked to go inside.

Sadie and Burt are chosen this time, leaving Luvie and Jessie in the compound. The two rejected chimps huddle together, whimpering with disappointment.

After a few minutes Luvie walks over to the door, picks up a stick from the ground and tries to open the lock with it. When this doesn't work, she gives the door a good whack with her fist and slouches back to Jessie. The two chimps hug each other again and wait unhappily for the others to return.

After the first series of tests, the testing is moved outside to the compound. Luvie and Burt are chosen this time, while Sadie and Jessie wait inside.

First Luvie is brought out into the compound by herself. She rides piggyback style on the tester's back with her arms wound around his neck and her legs wrapped around his waist. He takes her to a pile of rocks several yards off to the right of the door.

"Look, Luvie. See where I'm putting the banana," he says as he puts a piece of banana in between the rocks.

Luvie watches him carefully. Then they both go back to the door.

The object of this test is for her to get the piece of banana before Burt does. Burt doesn't know where it is hidden because he has been kept inside with the other chimps. However, if Luvie shows him in which direction the food is hidden, he will get there first because he can run faster. Luvie will have to fake him out or "lie" to him with her movements in order to get the banana.

Burt is brought into the compound and both chimps are released. They sit down by the door together and Luvie pretends to dig in the dirt with her fingers. Burt watches her, but Luvie doesn't look at him. Suddenly, Luvie bounds off to the right. Burt charges after her, passes her, and beats her to the pile of rocks, where he finds the piece of fruit. Luvie slumps back to the door, whimpering with frustration, while Burt sits munching contentedly on the banana.

During the next five tests Luvie loses the banana to Burt every time. Other tests show that all the chimps have difficulty in keeping the location of the food from their fellow chimps. They are unable to lie to each other about the hiding places.

After a morning of testing, the chimps are given an hour at noontime to play and relax before the afternoon testing begins. All the testers or trainers at the center take turns playing with them during this break, but Keith is the chimps' favorite. They walk with the others and ask to be swung around or tickled, but when Keith joins them they tumble all over him. They follow him to the tire swing, where he pushes them, sometimes two at a time. They wrestle on the ground with him, dig with sticks together in the soft dirt, share a soda with him, and gallop from one plaything to the next in a line strung out behind him.

At the poles the chimps scramble wildly up to the top, hang from the ropes, and drop to the ground in a heap. Then they dash off, chasing one another in a game of tag. They all make soft, grunting sounds to show their happiness.

Just like children, the chimps play nicely with each other for a while, then quarrel over a stone or other object, screaming and punching at each other. In a few minutes they are back together again as if nothing had happened.

Testing in the afternoon is followed by dinner. The chimps are brought inside to a big, wire-mesh cage for the night, and at five o'clock they usually start getting ready for bed. The lights are turned off at eight o'clock, but often the small, furry bodies are already curled up in sleep.

* * *

Burt, Sadie, Jessie, and Luvie are used in language research studies under the guidance of Dr. David Premack of the Psychology Department of the University of Pennsylvania. These tests are done at the Primate Center just west of Philadelphia.

The chimps are all about four to four and a half years old. They were captured in Africa when they were about a year old, and they came to the Primate Center six months later.

When they first arrived, the chimpanzees were isolated in a separate building for several weeks. This was to make sure they had no diseases that could spread to other chimpanzees.

During this time only Keith took care of them. He fed, cleaned, and played with them. They soon became

Keith shows Burt how to dig a hole with a sharp stone during their noon-hour play period

so dependent on him that he had to introduce all new trainers before the chimps would accept them. They seemed to feel that if Keith was with the new person, then he or she was all right.

The chimps are very different from each other in appearance and personality. They are easy to tell apart, once a new trainer gets to know them.

Burt, the only male, has black splotches on his cheeks. Sadie has a hole in her right ear. Luvie has markings on her eyebrows like a clown's makeup, and Jessie is the smallest, with the cutest face. Just like people's faces, their faces will change and mature as they grow older.

Burt is very obedient and pays more attention to his trainers. He is the most curious and brave, and, although he never bites people, he has bitten the other chimps. He is also the leader of the group, even though Sadie tries to be the boss, too.

The female chimps do bite people and will test new trainers by biting them to see what they will do. A good slap is the best way to warn them not to bite again.

Sadie and Jessie, who are the smartest of the four, always finish the problems the fastest. If they don't understand what to do, however, they will lose interest and not pay attention.

If Luvie gets a problem wrong, she will whimper, clasp her hands together, and press them against her forehead in great disgust.

The chimps have their favorites among the group

and will work better with them. Sadie and Burt get along well together and Luvie and Jessie are good friends, but Burt and Luvie don't like each other at all. They fight and bite each other and get very upset if left alone in a room together.

Occasionally, one of the chimps will escape from the cage and happily run through the lab, panting with excitement. The chimps are lightning fast and enjoy escaping, especially if it has been a boring day.

Their big adventure is usually a short one because they are so dependent on their trainers that they want to be with them. As punishment for escaping, the runaway chimp is put in a cage alone. Chimpanzees hate to be by themselves and will scream with rage when this happens. If scolded, they will cling to their trainer, looking for comfort and a few kind words.

The trainers also have to deal with temper tantrums. When angry or upset, the chimps will scream and roll around on the floor, kicking and hitting wildly at everything. A glass of water thrown on an angry little animal is the best way to stop the tantrum. Chimpanzees hate to get wet.

Most of the chimpanzees used in the study of communication are females. Although they are harder to work with and harder to control, they are thought to be smarter and do not grow to be as big or as strong as the males. A full-grown male can weigh up to 180 pounds and reach five feet in height, while a female is usually quite a bit smaller.

The chimpanzees used for the testing are usually under eight years of age. After this age, when they have begun developing physically into adults, chimps are even more wild and unpredictable. Because of their tremendous strength—three to five times stronger than a man of the same size—both males and females are usually too dangerous to work with when fully grown.

There have been many chimps raised in homes with people. Some of their foster parents even thought the baby chimps could be taught to talk. Except for a couple of chimps, including Viki, who learned to say three words—mama, papa, and cup—this proved impossible. A chimpanzee's vocal tract is not flexible enough for all the different sounds of human speech.

Visual languages, which depend on sight, are used now. Researchers found that they could use sign language, computers, and plastic symbols to "talk" with chimpanzees.

Washoe, a female chimpanzee born in Africa, was brought to the University of Nevada in 1966, when she was about one year old. Under the guidance of Drs. Beatrice and Allen Gardner, both psychologists, she was taught the same sign language used by deaf people. It is called Ameslan (American Sign Language).

As part of her training, Washoe lived in a house trailer that had a living room, kitchen, bedroom, and bathroom. She had her own toys, books, and games, and whenever she was awake there was a trainer present making signs to her.

Washoe learned four signs in her first seven months of training. They were "come-gimme," "more," "up," and "sweet." Her name in signs was "Big Ears" and she would brush her hands forward past her ears when talking about herself.

By the time she was three years old, Washoe used about 34 signs. At five she had a vocabulary of 132 signs plus a few others that were not part of Ameslan. She also understood several hundred more signs, but was still far behind the language skill of a five-year-old child.

Washoe thought that everyone knew sign language and even tried to "sign" with cats and dogs at first. She made her signs slowly and carefully with beginning students, and most of her signs could be understood by deaf people the first time they saw them.

Washoe had some of her own signs, too. They were considered slang and called "Washoese."

Other chimpanzees were taught signs from Washoe's vocabulary by Dr. Roger Fouts at the University of Oklahoma's Institute for Primate Studies. Sign language became such a natural part of life for one of the chimpanzees named Lucy, that she even tried to teach the sign for "book" to her pet cat. Holding the cat in her lap, Lucy patiently, but unsuccessfully, tried to help it make the sign with its paw while she pointed to the book.

Lana, a chimpanzee at the Yerkes Regional Primate

Center in Georgia, learned to communicate with a computer. Dr. Duane M. Rumbaugh of the Department of Psychology at Georgia State University was in charge of this study.

Lana lived in a special "learning environment." It was a seven-foot-square room with transparent plastic walls. On one wall were two machines that gave liquids, three that gave different foods, and two loudspeakers for broadcasting music. On another wall was a computer panel that looked like a typewriter keyboard. Each key was one inch high by one and a half inches wide. Each was marked with a design that stood for a word.

The "learning environment" was inside a larger room that had screens for movies and slide shows, a window so Lana could see outside, and a computer.

Lana could press the keys to "talk" to the computer or, through the computer, to "talk" to a person in the other room. She used her machine day and night to get food or drinks, or to entertain herself.

During the day she could press the keys to ask her trainer to come into her room and tickle her. At night she could ask for a sip of water, a movie, or a little music —with a choice of rock, jazz, or classical. In the early morning she could ask for a thirty-second peek outside by pressing on the keys to say, "Please machine make window open period."

Lana was born and raised in captivity. She was gentle with people who worked with her regularly and could easily be taken outside for walks. She was quiet for a chimp and didn't scream, as other chimps do.

During training Lana was never asked to do anything too difficult, but if she was bad, her trainer would leave the room. As a worse punishment he would turn off her keyboard. This last step brought pleading cries from the unhappy and very sorry chimpanzee.

Three trainers worked and played with Lana during the day. They groomed or tickled her, swung her by her arms or played ball with her.

When a teacher was in Lana's room, he punched out questions or suggestions for games on the chimp's own keyboard. Lana would then scamper across the room to answer his message. She was taught to start all requests with "please" and to end all sentences with a period.

Dr. Premack, who now does his testing at the Primate Center near Philadelphia, taught Sarah, an African-born chimp, how to "read" and "write" with plastic symbols. Each symbol stood for a word. The plastic pieces had metal backs so they would stick to a magnetized board. To earn a reward, Sarah had to put the right symbols in the right order on the board. She chose to place the symbols in a line under each other, so the sentences read from top to bottom.

Sarah worked with only one trainer during her lessons and spent the rest of the time in her cage. She and her trainer wore plastic symbols for their names on strings tied around their necks.

In addition to learning words, Sarah was taught some

of the rules of English grammar. The training program was divided into simple steps, each one leading to the next.

After five years of training, Sarah was able to use 130 symbols accurately about 80 percent of the time. Her vocabulary was about equal to that of a two-year-old person.

Today, Sarah, who is about fourteen years old, is at the Primate Center with Burt, Luvie, Sadie, and Jessie. She is being tested on problem-solving and must use her language to answer the questions. In addition to the plastic symbols, however, she also uses a pencil and a notebook now. It is amazing to see her turning the pages of her notebook and marking down the correct answers with her pencil.

One problem-solving test uses photographs. Four large color photographs may show four different ways to open a box of bananas that has been tied shut with string. Sarah must pick the photo showing the correct way to open the box. She may choose from photos showing a knife being used correctly to cut the string, a knife being used incorrectly (using the handle rather than the blade for cutting), a spoon, or a key being used to cut the string.

The purpose of teaching chimpanzees to communicate is not just to be able to "talk" to apes. It is hoped that doctors and other specialists can use these same methods with children who are unable to talk because of mental and emotional problems.

People may never really be able to talk to animals, but scientists involved in "monkey talk" are already finding ways to help people. Dr. Premack feels that Sarah's education and further work with chimpanzees can give valuable clues to brain function in people as well as apes. His system of dividing language into simple, step-by-step lessons has already been used to teach children and adults with brain damage. Dr. Rumbaugh and his associates believe that a computer set up for retarded people could be used to teach them to communicate or "talk" in the same way as Lana.

A mother of a retarded child asked, "If you can teach a monkey to talk, why can't you help my child?"

Drs. Premack, Rumbaugh, and others hope to have the answer soon.

7
Animal Actors

The big gray wolf crouched in the bushes and waited. A small herd of sheep could be seen grazing in the middle of the field. The wolf knew that they would head his way soon to escape the noonday heat under some nearby trees. Wolves usually hunt in packs, but this wolf was a loner and liked to be by himself.

As he watched, the sheep started toward him, paying attention only to the lush, green grass. The wolf singled out a sheep with a lamb at her side. They were grazing a little way from the rest of the herd and would be easy to catch off guard. The wolf's tongue flickered out, and he licked his lips at the sight of the plump little lamb.

The ewe and her lamb moved closer and closer. The wolf stood up slightly, with muscles tensed. He waited another second and then sprang from the bushes toward the two sheep. Bleating in terror, the ewe and her lamb turned to run, but it was too late. The wolf was almost on them.

Suddenly, a brown-and-white shape hurtled toward the wolf. It was Lassie—to the rescue.

The wolf stopped just before reaching the lamb and turned to face the collie. Lassie leaped at him, barking angrily.

The two animals reared up and then tumbled to the ground. They rolled over and over, biting and clawing at each other. Snarls and low-throated growls filled the air. The teeth of both dog and wolf flashed as they snapped at each other. Fangs sank into gray flesh, while claws pulled out hunks of brown-and-white hair.

The two animals separated and backed away from each other. Both dog and wolf were bleeding from their wounds. Carefully, slowly, they moved toward each other. Suddenly they sprang again, crashing together in midair before falling back down into the grass. The two seemed to be a whirling mass of gray, brown, and white.

Finally, they both retreated and stood staring at each other, as if feeling each other out. Both dog and wolf panted heavily. Again, Lassie approached the wolf, stalking stiff-legged toward him. Her hackles were up and her white teeth bared. Just as she lunged forward for another attack, the wolf turned tail and ran. He had had enough.

Lassie collapsed on the ground, too tired even to lick her wounds. She had done her job and protected the sheep, but it had been too much for her.

Just then a young boy appeared on the horizon.

"Lassie, Lassie."

The boy ran across the field to the dog and sat down

next to her. He put her head in his lap and petted her. He cried softly.

"You'll be all right, Lassie," he sobbed. "I know you will."

As the scene faded from the TV screen, the viewers too knew Lassie would be all right. She had to be. Next week was another show, and another story in the life of Lassie.

* * *

The above story is an imaginary description of one of Lassie's adventures. This famous collie, who made her first screen appearance in 1943, has always been played by male dogs. This is because their coats are more beautiful than females' and they are usually not so nervous or temperamental.

The seventh Lassie can now be seen in a new musical called *The Magic of Lassie.* Today he makes $150,000 a picture, while the first Lassie made only $200 a week.

Animal actors are very important to the movie industry. Some, like Lassie, have become stars, but not without hard work and long hours before the cameras.

There are more horses and dogs than any other animals in the movies—Gene Autry's Champion; Roy Rogers' Trigger; the Lone Ranger's Silver; Mr. Ed; Fury; Thunder; Rin-Tin-Tin; Sergeant Preston's malamute, King; Benji; and, of course, Lassie. But there have been other kinds of animal stars, too, such as Jake

Lassie is now a member of the Animal Hall of Fame

in *The Cat from Outer Space,* the lions in *Born Free,* Flipper, Francis the Talking Mule, Charlie the Lonesome Cougar, the bear who played in *Gentle Ben,* and Fred the cockatoo in the TV series *Baretta.*

At one time or another just about every kind of animal has been trained for movies—from rhinoceroses and giraffes to turkeys and cockroaches. One of the most unusual movies was *Willard* (1971), for which Hollywood animal expert Moe DiSesso trained 500 rats.

DiSesso started with twelve rats, allowed them to multiply, and then began to train them. He taught the rats to climb furniture, run in and out of a suitcase, climb stairs, bare their teeth, and squeak on cue.

He also taught them that people meant food and security and could be trusted. His son would lie down in the rats' cage fifteen times a day so they would get used to people. The rats were afraid at first, but soon they were sniffing their new cagemate and climbing all over him.

DiSesso trained the baby rats to identify the sound of an electric buzzer with food. He rewarded them with dabs of peanut butter. By the time the rats were adults, they took directions and worked well together in groups of twenty-five.

Animals that are chosen to be stars must have several qualities. For horses, a major requirement is beauty. They must be well formed, have expressive eyes, spirit, and good color. Many of the horse stars are American saddlebreds. Their flashy way of moving, their flowing

The Lone Ranger's horse, Silver, was a favorite among TV viewers

tails and high-carried heads make them stand out from other horses.

In the early days of movies, the good horse was usually a white stallion, while the bad horse was a black stallion. Today, black is a favorite color of actors, as is white, but a rich chestnut color is popular, too. One handsome chestnut horse with a white blaze on his face and a long, silky mane and tail has been the favorite of several actors, including John Wayne, for many years. He has been ridden in a number of Westerns.

Horse stars must have above-average intelligence, be obedient, be able to accept hard work, have trust in people, and be able to work in spite of the noise, confusion, and strange machines on a movie set. Statistics show that only one horse out of five hundred has all the qualities to be a star.

When choosing a dog to act in a movie, animal trainer Carl Miller looks for a good appearance with as light a color as possible, an outgoing personality, and spunkiness. A solid black or a solid white dog is not a good choice for movies because such a dog produces a flat photograph with no depth to it.

Although the German shepherd has been used more than any other breed of dog, many of the dogs in today's pictures are mixed breeds. Mid-America likes the mutt and can easily identify with it, because many people have and love mutts of their own.

The cats used in movies have to like people and want to be with them. Most cats like to be left alone, so are not good for movie work. A cat must be eager to show off and must not be nervous or afraid of loud noises and unfamiliar surroundings.

A large number of canine movie stars have been found in local animal shelters. Benji, the little floppy-eared dog with sad eyes, was found in the Burbank, California, animal shelter by his trainer, Frank Inn. He is said to be a blend of cocker spaniel, poodle, and miniature schnauzer. Benji's ability to learn quickly, his eagerness to please, and his feelings and responses to

Benji captures audiences with his expressive face and sorrowful eyes

the different emotions of his trainer—happiness, sorrow, anger—are what make him such a good animal actor.

Benji, who is now a teen-ager, shares his work load with Benji, Jr.

Spike, star of the Disney film *Old Yeller,* was found by dog trainer Frank Weatherwax in the Van Nuys, California, animal shelter. He cost $3, but Weatherwax invested four years and $20,000 worth of training in him before he played in his first movie.

Sandy, the scruffy but huggable dog in the Broadway musical *Annie,* was found by his owner in a dog pound in Newington, Connecticut, and purchased for $8. Now

Sandy gets fan mail, advice, and presents from his many admirers. After performing in Washington, D.C., at the Kennedy Center, he even shook hands with the President of the United States.

Four members of the Doberman Gang were also dogs without homes. Three came from one family that

In front of the camera, Benji waits for a signal from his owner, Frank Inn, before making a move

was going to be evicted from their home because the dogs barked too much. Another had been left at a boarding kennel and was never picked up. With the owners' permission, the kennel sold the dog to Carl Miller to cover the boarding expenses.

Toto, the little dog that stars in the movie *The Wiz* with Diana Ross, was brought to a New Jersey veterinarian to be put to sleep if nobody wanted him, because he was so wild and unmanageable. The vet didn't want to put the healthy, eleven-month-old schnauzer puppy to sleep, so he called animal trainer Leonard Brooke, who trained the puppy and made him a star.

Many of the horses and other animals seen on movie screens were once unwanted or came from places not connected with Hollywood.

Misty, a sleek, almost black thoroughbred stallion, was picked up at a racetrack after a leg injury ended his racing career. He played the villain for almost twenty-five years and appeared in about seventy different pictures. His best role was as a "fighting stallion." Although a real gentleman behind the scenes, he would snake out his neck and show his teeth when told to, just like a real outlaw stallion.

Fury, one of the biggest money earners in Hollywood's animal kingdom, was found on a farm in Missouri when he was a year and a half old. The black stallion learned to open doors, run to the schoolyard to pick up his young owner, and poke his head into windows. He even allowed a gang of motorcyclists to chase him down a highway during one of his pictures.

Cocaine, another Hollywood horse, was on his way to the slaughterhouse because of a severe leg wound. He was rescued and became one of the best stunt horses in the business.

Bozo, the female grizzly bear who plays Ben in the TV series *The Life and Times of Grizzly Adams,* was found in a circus. In spite of the reputation of grizzly bears as the most ferocious and untrainable of all animals, Bozo works freely without a leash. Marshmallows are her reward for performing tricks correctly, and in one season she ate so many that she gained 106 pounds, ballooning from 501 to 607 pounds.

Picking animals for a movie sometimes depends on the type of story to be filmed. A light comedy might have a skunk, monkey, or clumsy Great Dane, as in Disney's *The Ugly Dachshund,* while a dangerous, scary movie may have a lion, snake, bear, or shark, as in one of the biggest moneymakers of all times, *Jaws.*

The more unusual animals in films, such as chimpanzees, lions, birds, bears, and deer, must look good, but their ability to work well with people is even more important. An animal that can't be trusted around people is too dangerous to have on a movie set if it has to work alongside actors without muzzle and chain.

Wild animals raised by knowledgeable animal trainers usually learn to trust people and are good at taking directions. Clarence, the cross-eyed lion in the TV series *Daktari,* was such an animal. He worked well off lead and was gentle with all his fellow actors, whether they were two-legged or four-legged. The animal actor most unpopular with movie people is the chimpanzee, because it is so dirty and may bite suddenly for no reason at all.

An animal being trained for the movies spends his first weeks just getting used to his trainer and to his new home. Dogs are usually taught the basic obedience commands, but after that they start learning whatever stunts are needed for a particular picture. During this training they are also introduced to movie sets so they will learn to work in spite of flashing lights, noise, busy people hurrying everywhere, cables winding across the floor or ground, and cameras rolling from one spot to another.

A movie set can be especially frightening to horses and wild animals who are used to the quiet of a barn or cage. The lights that flash in their eyes and the noise and heat are unpleasant, but even more disturbing are the cables that look like snakes and the rolling cameras that look like monsters ready to pounce on them.

Training an animal requires patience and repetition. A trick must be worked on several times a day for short periods of time so the animal learns it, but doesn't get bored. All animal training is based on love, respect, and discipline.

Dogs need plenty of praise, while horses succeed with food rewards for a good performance and physical correction (a jerk on the halter, a smack on the neck, or a crack of the whip) for misbehaving. Other animals, such as cats, monkeys, raccoons, lions, bears, and dolphins, learn to do tricks with food rewards. At mealtimes, they get only half their daily portion of food. The rest is given to them throughout the day as rewards during training or filming. A well-trained animal is happy and not scared or forced into doing a trick.

Some of the stunts taught to animals are so unbelievable that moviegoers often feel they must be faked. DiSesso trained his dog, Bracken, one of several that play the bionic dog for the TV series *The Bionic Woman,* to chase a runaway Volkswagen, grab it by the bumper and stop it.

He has cats that turn lights on and off, dogs that open a door by grabbing the knob and pulling on it, a raccoon that plays basketball, a turkey that rings bells, hens that bang away on a typewriter, and a goose that plays the drums.

One of DiSesso's longest stunts involved Rennie. As the "bionic dog," Rennie chewed a lock off his cage, pushed the cage door open, walked over to an outside door, opened it and walked out—all in one take without any mistakes.

Scruffy, the little dog in the TV series *The Ghost and Mrs. Muir,* was taught by Miller, his trainer, to play the piano, turn lamps on and off, and pick up papers from the floor and put them in the wastebasket. One of his most unusual stunts in the series was chasing actor Charles Nelson Reilly, who played the landlord. Out the door and down the front walk they ran, but as Scruffy went through the open gate he did a backflip, slamming the gate shut with his hind feet.

Miller has a basset hound who picks up its own ear and holds it in its mouth, another dog who hides its face in a fishbowl, and a cat who flips through a book as if he is reading it.

Horses used in a leading role usually go through a three-month training program. They learn to walk toward the camera, stop when told to, back up, look to the right or left, limp, and play dead on cue. A "dead

Mr. Ed takes a bite of a fellow actor's salad during the filming of his television series

horse" must not move when the actors are saying their lines, because one flick of an ear will ruin the scene.

Mr. Ed, the golden palomino in the TV series, learned to pick up a telephone, open and close his stable door, hold a large pencil in his mouth, pick up a hat, wave a flag, and pick up coins.

Gene Autry used several horses to play Champion, but the most well-known was a Tennessee Walking Horse. He was trained to kneel, bow, do dance steps, smile, kiss, and leap through a large paper poster.

Many horses in old-time Westerns were taught to untie simple knots with their lips in order to free their masters from Indians or bad men. Others are still being trained to fall down when their heads are pulled to the right and back. Most stunt men train their own horses to do this. It takes a long time and must be done correctly without hurting the horse or he will refuse to do it. The ground is dug up in the area where the horse is to fall. It is replaced with ten inches of sawdust to cushion the horse and keep him from being hurt.

All animal stars have doubles or stand-ins. To keep the star fresh for close-ups, the doubles may do the running scenes or other activities filmed from a distance. They usually take the star's place on a movie set while the lights and cameras are being set up.

If a double is used in the film, it must look exactly like the star. Makeup often has to be used to make the two animals look alike. Spray paint may be used to darken the paw of a dog or cat, while a clown's white makeup may be used to make a star or a blaze on a horse's face.

Makeup is also used to change an animal's appearance for the role he plays. Country Gentleman, the horse that played in *Smokey,* was supposed to look starved and mistreated. Makeup men painted shadows on his haunches so it looked as if his bones were sticking out. They also drew gray lines around his eyes to make him look dull and beaten.

Movies have always been able to make things look different than they really are. The fight scenes between an animal and a man or between two animals, such as the one described earlier between Lassie and a wolf, are good examples of this. Using trick photography, camera angling, and cutting and splicing film can easily make it look as if the opponents are trying to kill each other.

For a fight scene the two animals are muzzled with thin wire, tape, or a shoestring to keep them from biting each other. The muzzles don't show, because the action is happening so fast. Horses have rubber boots slipped over their hoofs so if one gets kicked, he won't be injured. For the Lassie fight scenes, close-ups of first the dog and then the wolf are shot, showing them snarling and snapping at each other. Actually, they were taken one at a time and told to "watch" a stranger who was pretending to attack their master. The snarling and snapping was filmed and then put between other shots of the dog and the wolf "fighting" so it looked as if they were trying to get at each other.

Close-ups of Lassie biting into the wolf are also put in, although the dog is really biting into an old, gray animal hide. When the wolf pulls brown-and-white hair from Lassie, it is really a close-up of him clawing a piece of fake brown-and-white fur.

If a horse is supposed to kick a dog for a movie, the first scene is of the horse kicking out. The second scene shows either the horse's hoofs or a mechanical set of legs and hoofs kicking a stuffed dog. The third scene shows the trained dog tumbling and rolling over as if he has been kicked. Put them all together and it looks as if the horse is kicking the dog.

If a horse is fighting a snake, film editors use close-ups of a real snake, close-ups of a horse's head, and then close-ups of fake horse legs and hoofs crushing a mechanical snake.

In *Gunsmoke,* another TV series, a group of dogs, according to the story, were trained to protect their outlaw masters. At the end of the show, one dog was clubbed to death and another was drowned.

In real life, the dog that was "clubbed" was actually ten feet away from the man with the club (a rubber one). The dog was told to "play dead." Then, with cameras rolling, the dog was told to jump up, just as the man swung the club in the background. The film was reversed, refilmed in reverse, and spliced into the final film. To viewers it looked as if the man swung his club and hit the leaping dog, who fell dead on the ground.

The dog who was "drowned" was filmed jumping up

on a horse and knocking the rider out of the saddle. The rider and the dog rolled down a hill into the water. Then the film was cut and the dog left the scene. An air gun was put under the water and the filming started again. The rider acted as if he were holding the dog underwater. All the audience saw was air bubbles, supposedly from the dog's lungs, foaming up to the water's surface.

These scenes looked so real that thousands of letters poured into the studios complaining of cruelty to the dogs.

In the early days of movies there was much cruelty done to animals. Horses were tripped, rather than trained to fall down, and often suffered broken legs or necks. Animals really did fight in fight scenes, and often needed a veterinarian to sew up their wounds. Many worked for hours on hot sets without a break—only to come back the next day and do it all over again because some human actor had messed up his lines.

Today, animal actors have their own protection society—the American Humane Association. In the United States, the AHA always sends a representative to movie sets that are using animal actors. They check up on animal health, comfort, and safety, and they offer advice on care and handling.

In the 1925 filming of *Ben Hur,* about 150 horses were killed during the chariot races. In the 1959 version not one animal was even hurt.

The AHA has been responsible for other improvements, too. During the filming of *Smokey,* Country Gentleman, who played the lead, was given a two-hour lunch break, while the human actors received only half an hour.

Horses who kick down barn doors are actually kicking apart a door made from balsa wood—a very soft wood that breaks apart easily.

A covered wagon that tips over while being chased by Indians is very dangerous to the galloping horses that are pulling it. The AHA designed a device that releases the horses from the wagon seconds before it turns over.

In 1951, the AHA started an annual awards ceremony for movie animals to show that Hollywood animals are well treated and properly trained for their roles. The Patsy, which stands for either Picture Animal Top Star of the Year or Performing Animal Television Star of the Year, is equal to the Oscar and Tony Awards given to human actors.

The Patsy recognizes the talents of the animals. However, it is given only to animals who receive good treatment from their trainers and from the studio filming the picture. Famous movie and TV stars hand out the awards during the ceremony.

The Patsy now has four different categories—a canine division, an equine division, a wild animal division, and a special division for all other animals. Francis the Talking Mule won the first Patsy for his comedy performances in 1951. Other winners have been Lassie; Caesar, a Doberman pinscher who starred in the telemovie *Trapped;* Benji; Scruffy; Pax, the white German shepherd who played a blind detective's guide dog in the TV series *Longstreet;* Fred the cockatoo; Arnold the pig in TV's *Green Acres;* and Esmeralda, the seal in *20,000 Leagues Under the Sea.*

Lassie, Benji, Francis, Scruffy, and Morris are not allowed to compete for the Patsy anymore. They have won the award so many times that they have been placed in a special Hall of Fame so that other animals will have a chance to win, too.

Hollywood movie studios have learned that audiences love stories about animals. Some of the all-time favorites—*Black Beauty, My Friend Flicka, National Velvet, Lassie Come Home, The Yearling, Born Free, The Day of the Dolphin, Benji, Old Yeller,* and *The Call of the Wild*—have brought millions of people of all ages into movie theaters. Animal actors rarely hear applause for their accomplishments. Instead, they work for peanuts . . . or fish . . . or oats . . . or a bone . . . or just a pat on the head. That's usually all they want.

8
Mule Express

The two mules stood patiently at the hitching rack beside the canal. Already harnessed, they were waiting to start their day's work. Occasionally, one twitched a long ear or flicked a tail to chase a fly away.

Lissi, a large dark-brown mule, closed her eyes and began to doze. She was the leader and stood in front of Spooky. Spooky was a pretty chestnut color with blond mane and tail and a thin white blaze down her face. Both mules were still a little shaggy-looking because they hadn't completely shed their winter coats yet.

Spooky stamped a foot and looked back to see what was taking so long. She was tired of waiting. Just then a tasty-looking bush to her right caught her eye. Its tender new leaves had just opened wide. Spooky stretched her neck out as far as the rope and her halter would let her and began to nibble them.

A colorful green-and-yellow barge floated in the canal several yards behind the two mules. Next to it was a line of children and their mothers waiting to climb on board the large, flat-bottomed boat. Finally everything was ready and the group began, one at a time, to step onto the barge deck. They all found seats under the fluttering green-and-yellow-striped canopy. The children quickly chose the benches along the sides, while their mothers sat in the deck chairs in the middle of the barge.

It was a warm, spring afternoon and the children chattered excitedly about their trip. They had been studying Pennsylvania history in class, so they knew all about the mule-drawn barges that plied the Delaware Canal during the 1800's. Now they would be able to see for themselves what it was really like to ride on one.

While the passengers were settling themselves on

Mr. Schweikhardt hitches Lissi and Spooky up to the barge's towline

board the barge, Jurgen Schweikhardt, owner of the barge operation, hooked one end of the long towrope to the front of the barge. Then he carried the other end down to the mules.

Carefully, he backed the mules away from the hitching rail and stood them, one behind the other, on the towpath. Next he attached the towrope to the harnesses. Spooky stood slightly to one side of Lissi with her head next to Lissi's left hip. The towrope split in a "Y" just behind the mules, so that one end was hitched to Lissi's harness and the other to Spooky's. This way both mules pulled the same amount of weight.

Lou, the driver, came up and took Lissi's lead line in his hand. He would walk alongside her to guide her down the towpath.

Jurgen Schweikhardt untied the barge and jumped on board. He stood at the back of the boat, with a hand on the tiller to steer it.

"O.K., let's go," he called out.

"Come on, girls, pull," said Lou as he gave a tug on Lissi's halter.

Lissi and Spooky leaned into their harnesses and slowly moved forward. One step at a time, muscles straining, the two mules pulled together.

After moving a few feet, the barge floated easily and the mules relaxed. They had no trouble walking now. Once started, the barge just needed a light, steady pull to keep it gliding down the canal.

The barge made a swishing sound as it passed

through the muddy, brown water. Bushes and tree branches hung out over the opposite or berm side of the canal. The dirt towpath was packed down hard, showing its many years of use.

The children continued to chatter, pointing out to each other the various things they saw.

"Look how fast we're going," said one little girl.

"Hey, did you see that snake?" asked a boy, leaning over the side.

"Keep your hands in," warned his mother.

"Oh-h, look at the baby ducks!"

Eight little brown balls of fluff floated in a line behind their mother. As the barge approached, they crowded around her. She herded them into a little cove between a rock and an overhanging tree branch and waited for the barge to pass.

Lissi and Spooky continued to plod along the towpath. The bells on their harnesses tinkled with every step. Lou had hopped up on Lissi's back now and was enjoying the ride. He got off and walked whenever they went under a bridge.

In town, houses were built right up against the canal, leaving room for only the narrow towpath. The barge floated past both tiny old crooked buildings built in the last century and modern wood-and-glass houses. The mules walked under camelback bridges, past dogs sniffing delights in the grass, past people relaxing on the benches, and next to joggers and strollers taking advantage of the good weather.

The canal left the town and continued on between fields of tall grasses and wild flowers. An occasional weeping willow hung out over the path and the water, giving both mules and passengers a moment's relief from the hot sun.

At the halfway point of the ride, the mules were reversed and led past the barge, now stopped and float-

Lou leads the two mules under one of the low camelback canal bridges

ing next to the bank. Jurgen Schweikhardt switched the towrope and tiller around so that now the front of the barge was the back and the back was the front. A towpath is always on one side of a canal only. Mules must pass on this narrow path.

"Move out," he called.

Lou gave a gentle tug on Lissi's halter, and the two mules leaned into their harnesses, starting back down the path on which they had come. Soon the barge floated easily again, retracing the first half of the ride.

The trip back to the loading area was just as peaceful as the one going away from it—past dogs, strollers, joggers, and baby ducks. Back at the loading area Lissi and Spooky lowered their heads, closed their eyes and took a nap until the next trip. Today it is easy work for a mule to pull the canal boats. Long ago it was a tiresome, more strenuous job.

* * *

Lissi and Spooky are owned by the mule barge operation in New Hope, Pennsylvania. They are only two of the eight mules that haul passenger barges along the Delaware Canal.

In the past there were hundreds of mules plodding up and down this hard dirt towpath. They pulled canal-boats full of coal, iron, logs, lumber, grain, flour, lime, limestone, and other products from Easton, Pennsylvania, to Bristol, Pennsylvania. Then they hauled the empty boats back up to Easton—120 miles round trip.

It was slow going. A loaded boat moved about two miles per hour, while an empty one was able to go only about four miles per hour.

The days were long for the teams of mules back in the 1800's. On the Erie Canal in New York State, traffic continued day and night, but the boats were large enough to carry an extra team and driver. Every six hours the teams switched, and the team and driver that had been resting took over for the ones on the towpath.

On the Delaware Canal there was no room on the barges for an extra team and driver, so the workday was eighteen hours long!

Mules were used rather than horses because they lived longer lives, were more surefooted, and could work the long hours without becoming sick. Horses can walk faster than mules, but they can't keep it up for as long a time. Donkeys were too small to pull the huge loads.

In addition to being bred on local farms, canalboat mules were brought to the area from Kentucky and Missouri. The Kentucky mules were said to be fancier-looking, but they did not hold up as well as the Missouri mules.

When buying a mule, the barge operators looked for good legs, strong shoulders, and broad, muscular hips. Some of the mules were pretty wild and had only learned to wear a halter. After the mules were broken to the harness, they had shoes put on their feet. Then a number was branded into one front hoof with a branding iron. Each number was recorded under the buyer's name.

Most boatmen were proud of their mules and harnesses. They liked healthy, good-looking mules and fancy harnesses. Some of the harnesses had brass rosettes, brass bands, and other ornaments. Almost all the mules wore bells—two on each mule. They were attached to a strap that went from the underside of the collar down between the front legs to the belly band. The ringing of the bells helped the captains steer the barges during foggy weather.

The bit could be unstrapped from the bridle at one end so it could be taken out of the mule's mouth while he was eating. Many of the mules ate their meals from feed baskets tied over their muzzles as they walked along the towpath.

The mules wore waterproof blankets to cover their backs and shoulders during rainy days, and fly nets to keep off the flies in the summer. Some mules even wore straw hats to shade their heads from the hot sun. The hats had two holes cut out for the ears and a band that was tied under the chin.

Representatives from the SPCA (Society for the Prevention of Cruelty to Animals) traveled up and down the Delaware Canal checking on the condition of the mules. If a mule was not cared for properly and had a sore on his back or shoulder where the harness rubbed, the owner had to take his animal out of harness immediately, give him rest and care, and pay a fine.

After a long day's work, most of the mules liked to lie down and roll to scratch their backs as soon as their harness was taken off. A "roller" was a mule who rolled whenever he liked, even if he still had his harness on.

Some of the mules were trained to plod along the towpath by themselves so the drivers could get on the boat for a meal with the captain and a short nap. Lumps of coal or stones were thrown at the mules if they stopped to eat grass. But a mule had to be given water regularly, or he might walk right into the canal for a drink.

Often the drivers who guided the mules were small boys, eight to ten years old. The days were pleasant for them, but the hours spent in the dark in the early morning and the evening were cold and scary to the youngsters. The boys would snuggle up to the mules' sides for warmth and companionship.

At night the mules were kept in stables, which were usually built on the opposite side of the canal from the towpath. This far side was called the berm side. The mules would have to cross over the canal on a bridge to get to them. Most captains had their favorite stables along the route and stopped at them regularly.

The mules were cleaned, harnessed, and on the towpath ready to move by 4 A.M. Some owners gave their mules a light breakfast of oats and corn while they were cleaning and harnessing them. Others waited for about an hour after they had started working and put feed baskets on them so the mules could eat and walk at the same time. The mules were fed four times a day and given plenty of hay at night.

Most of the teams numbered two mules, but sometimes there were three and even four in a team. The mule in front was called the lead mule and the one in the rear was called the shafter.

The towpath ran along only one side of the canal. When boats going in opposite directions had to pass, the empty boat had the right of way and would move to the towpath side of the canal. The loaded boat would steer to the far or berm side. Its mules would move off the towpath on the side away from the canal and stop. The towline would sink to the bottom of the canal and the approaching mules and boat could step or pass over the towline and be on their way.

During the winter when the canal froze over, the mules were put out to pasture and the canalboats hauled out of the water so any repairs could be made.

In the spring, the mules were brushed and cleaned. New shoes were put on their feet and the harnesses were scrubbed, oiled, and polished.

The Delaware Canal was almost a solid mass of boats during its busiest years in the middle 1800's. Pennsylvania, Indiana, Ohio, and New York had many miles of canals. The Erie Canal, opened in 1825, was 364 miles long, reaching from Albany, New York, to Buffalo, New York. There was even a song written about it.

The end of the canals finally came when the railroads

took over. In the spring of 1931, there were only twenty boats in operation on the Delaware Canal. That fall, traffic on the canal stopped for good.

Mule-drawn boats are still used on canals in Europe, but in the United States there are only pleasure barges in operation on the canals. In addition to the one at New Hope on the Delaware Canal in Pennsylvania, there are at least three other barge rides—in Georgetown near Washington, D.C., on the Juniata Division of the Main Line Canal in Lewistown, Pennsylvania, and on a section of the old Delaware and Hudson Canal in White Hills, Pennsylvania.

Today's mules are picked up from farmers, auctions, or horse dealers. Jurgen Schweikhardt looks for mules that have been broken to the harness, aren't too old, and aren't too tall. If they are too tall, they won't fit under the canal's low bridges. One of the mules who tossed her head and hit it while going under a bridge last summer refused to walk under a bridge again for several months.

A well-trained, healthy mule costs between $500 and $700.

The New Hope barge operation has five female mules and three males. The females are usually gentler and calmer. None of the mules bite or kick people, but they will let loose on one of their fellow mules if annoyed.

The eight mules are divided into leaders and shafters. The leaders usually walk faster, are more experienced, and are less likely to stop and graze.

The mules work from April until the end of October, depending on the weather. On Friday and Saturday evenings they may make four trips, at 6 P.M., 8 P.M., 10 P.M., and 12 P.M.

The night trips are a lot of fun for the passengers, who sing, laugh, and call to the people they pass while going through town. Many of the day trips are taken by groups of schoolchildren.

A trip on one of these mule-drawn barges gives passengers a peek at what life was like along the canals in the 1800's. It's a pleasant way to spend an hour learning about a part of our country's past.

9
Lifesaving Newfs

The beach was crowded on this Labor Day Weekend. Saturday and Sunday had been clear and sunny, but Monday proved to be even hotter and brighter.

The sun continued to climb in the sky as the people —singles, couples, and whole families—spread their blankets and lathered up with suntan lotions and oils. Here and there, an ambitious person speared the sand with a beach umbrella. Others just accepted the heat and lay back to soak up the sun's rays.

It wasn't long before the water's edge was crowded with splashing toddlers, castle-building youngsters, and nonswimming waders. Farther out, those trying to ride the waves shared the deeper water with the drifters on blow-up floats.

A young man watched over the mass of people from his seat high up in the lifeguard's chair. At the foot of the chair sprawled a big, black Newfoundland resting in the shade of an umbrella. The dog had dug a shallow hole so he could lie in cool, slightly damp sand rather than on the burning layer on top.

Caesar was an old hand at beach life. He only lifted his head when Don, his master, blew the whistle to warn those swimming out too far to come in closer to shore. Theirs had been a safe, uneventful summer so far, but one never knew when saving a life was going to be necessary. Both Don and Caesar were always ready for an emergency.

The afternoon continued without problems.

It's going to be another easy day, Don thought to himself. Then he noticed a young boy, about twelve, on a small raft floating quite a way from shore. Don whistled and motioned him to come in, but the boy didn't seem to hear or see him.

A couple of hundred yards from shore two large waves began to form. They rolled across the ocean's surface, gathering strength and size as they got closer to land. Don could see that they were going to break farther out than the other waves had, probably right about where the boy on the raft was drifting. He stood up and frantically began waving his arms and blowing his whistle. The boy still didn't look up.

The first wave rolled under the raft, making it rock back and forth violently. The boy looked up in surprise. Then he saw the second wave towering over him, but it was too late. As Don watched, horrified, the wave crashed down on the boy, sending his raft spinning off toward the shore.

"Caesar, up."

The dog sprang to his feet and looked up. Don tossed

a life ring down to him.

"Take it, Caesar," Don commanded.

Tied to the life ring was a three-foot line with a knot at the other end. Caesar grabbed the knot, held it in his mouth, and looked up at Don again.

"Caesar, go out."

Don motioned with his arm. Following the motion, Caesar galloped toward the water off to the right of the lifeguard's chair. The life ring bounced along behind him, but the knot was clutched tightly between his teeth.

The big dog bounded into the surf, leaping high and splashing down until the water was deep enough for him to swim. Other swimmers scurried out of his way, but Caesar never looked at them. He plowed forward through the waves, towing the life ring behind him.

Don could see the boy's head now, bobbing up and down between the waves. The raft had been carried several yards away. The boy would have a long swim to reach it.

As Don watched, the boy began to tire. He had swallowed salt water when the second wave had crashed over him. Because of the rolling surf, he couldn't see his raft. Scared and confused, he yelled for help.

His cries were heard by other swimmers, who looked up to see what was happening. Don could see Caesar's big black head pushing through the water. The life ring floated behind him.

The current pushed the boy more to the right. Don

gave three short, sharp blasts on his whistle. When he heard the signal, Caesar turned to the right slightly and then continued forward.

Finally, Caesar saw his victim. The boy was very frightened and just barely able to keep his head above the water. Caesar headed straight for him. Just before the dog reached the boy, he turned to the right to circle in front of him. The life ring swung out within the boy's reach. The boy lunged forward and grabbed the edge of the ring. Caesar came on around and headed for the shore. It was slower going now because of the boy's weight, but the dog paddled forward with powerful strokes.

Don watched for a minute to make sure that Caesar was all right. Then he jumped down from the chair, grabbed the first-aid kit, and ran to the water's edge. The other people had gathered to watch the rescue. As the dog and his victim came in closer, Don and a couple of other men waded out to get them.

"Come on, Caesar," called Don. Then, "Good boy," as he grabbed the child and carried him up on the beach.

"Someone bring a blanket, quick."

A woman snatched her blanket from the sand and ran it over to Don, who wrapped the boy up. As Don began to check the boy, he thanked God for having Caesar with him on the beach that day.

Caesar pulled the life ring up onto the shore and gave a couple of good shakes that sent a spray of water over everyone within six feet. Then he settled down to wait for Don. He knew he had done a good job, but it was all in a day's work for him.

* * *

This story about Caesar is imaginary. But it is typical of the work of a lifesaving Newfoundland.

Although there are canine lifeguards already on the job in other parts of the world—Europe and Australia—there are few, if any, in the United States. But many breeders and owners of Newfoundlands all across the country are trying to do something about that. They are training the dogs for water rescue work and hope they will be stationed on beaches in the future.

Newfoundlands are specially equipped for working in the water. They have webbed front paws to help them swim well, and two coats of hair to keep their skins warm and dry, even after long periods in the water. The top coat is straight and rough, while the undercoat is very soft and thick.

Newfoundlands are large dogs. The males may weigh 140 to 150 pounds and the females 110 to 120 pounds. They have broad, square heads, strong necks and backs, deep chests, and muscular hindquarters. Their most common color is a dull jet black, possibly with some white on the chest and toes, but they can also be a chocolate color or white with black markings. A white Newfoundland with black markings is called a Landseer, after Sir Edwin Landseer, a well-known artist of the 1800's who included the black-and-white dogs in many of his paintings.

The Newf is a gentle, intelligent dog with a sweet disposition. They are devoted companions and, throughout their history, have been working for their masters. Although also used for pulling small carts, the Newfs' lifesaving instincts have made them particularly valuable in or near the water.

They played a major part in helping to develop the island of Newfoundland, after which they were named, helping the island's fishermen pull in their nets. As the fishing industry grew, sailors and dock workers became more aware of the dog's ability to swim. Soon Newfoundlands were going around the world on commercial sailing ships. They carried the ships' towlines ashore and were always ready to rescue men and retrieve equipment that fell overboard.

There are many stories about rescues by Newfoundlands. They have pulled people out of dikes in Holland, from stormy seas off Canada, and from bathing areas of both Europe and North America. Newfoundlands also have rescued other dogs. They have been known to dive for fish. Their breed may even have been used by the ancient Romans. In the museum at Naples, Italy, there is an antique bronze statue from the Roman era showing two large dogs dragging drowning persons from the sea.

A Newfoundland is buried on the grounds of Windsor

Sebastian, a Landseer Newfoundland, inspects a boat bumper, then demonstrates an exercise in the Senior Division Water Tests and finds an object hidden under water. He tows a life ring to a drowning "victim"

Sebastian brings his "victim" back to shore. A good toweling helps keep the family station wagon dry

Castle in England because he rescued a drowning man, and, if it hadn't been for one of these large, gentle dogs, Napoleon might also have been lost at sea. During his first escape from the island of Elba—where he was sent and held for about a year—Napoleon fell overboard in the dark. A Newfoundland, belonging to one of the sailors who had come to free the military leader, jumped into the sea and pulled him to safety.

In 1919, a ship got stuck on the rocks off Bonne Bay, Newfoundland. The ship's dog, Tang, swam to shore with a line, so that all 91 passengers and crew members could be hauled to safety. The waves were too big for a boat to get close enough to the stranded ship, but the dog made it in spite of the rough surf.

France has made good use of Newfoundlands. Teams of them were once kept on the banks of the Seine River in Paris to pull drowning people out of the water. Others have been trained to work in pairs and are presently stationed on some of the major beaches in Brittany, France. Working together, they have been taught to flip a drowning person over onto his back and pull him ashore by grabbing on to his shoulder or upper arm.

One Newf recently rescued a small boy who fell through the ice while skating on Long Island Sound. The dog held the boy above water until help came.

Charcoal, another Newfoundland, was credited with saving two children from drowning during the summer of 1951 at Southampton, New York. At first the dog frightened many people because of his size. He annoyed others because he was always paddling alongside them as they swam. However, after the two rescues he was not only allowed on the beach but encouraged to stay there.

The Newfoundland Club of America started water tests about seven years ago to keep the breed's ability for water rescue work from being lost. The water tests are divided into a junior division and a senior division. They combine obedience training with a testing of the dog's natural instinct for water rescues.

The first exercise in the Junior Division simply asks the dog to swim with its handler. It isn't as easy as it sounds. Not all dogs—even some Newfs—are eager to go in the water.

Some of the other exercises in this division include retrieving a boat bumper, taking a 50-foot line out to a boat or a person, and towing a boat along the shore for 50 to 75 feet while wading in shallow water. If the dog fails any one of the exercises, he fails the whole test.

In order to try to pass the Senior Division tests, the dog must have first passed the Junior Division. Some of the exercises included in the Senior Division are jumping into the water from a dock and retrieving a boat bumper, retrieving two items floating 50 feet apart while the handler directs the dog to each one, towing a boat to shore, taking a life ring out to a boat, and retrieving an object on the bottom about 1½ to 2 feet underwater.

Newfs usually show a love of retrieving, swimming, and lifesaving. This love must be combined with carful training if the dog is to be successful in water rescue work.

A young puppy should not be overtrained in the water, but should be allowed to splash around with the family so he becomes used to it. Obedience training—heel, sit, stay, down, come, and stand—can be taught during this first year. It will be the foundation for later training in water rescue.

Puppies are given the command "Take it" whenever they are given a piece of food or a toy. To teach them to go after something in the water, a dog biscuit is floated in a large water dish and the dog told to "take it" from there. Next the dog has to retrieve the biscuit while the trainer holds it partly underwater with one finger. Finally, the dog is told to "Take it" when the biscuit is held on the bottom of the dish completely underwater. This way they learn to take whatever is shown them on command, building up to a boat bumper, towline, and life ring, without being afraid of getting their faces wet.

Running along the water's edge with a Newf, and playing with him in shallow water helps the trainer introduce the dog to swimming. Another dog swimming with them encourages the young dog to swim, too.

The dog will think the training is fun at first, especially when splashing around in the water, but he must learn to do what he is told. Water training is begun with a long lead so the dog can be made to do the exercises correctly and not in his own way or at his own pace. This discourages the dog from refusing to work if he ever decides the training is not fun anymore.

A firm "No" is usually all that is needed to tell the dog he is not doing something right.

A third division of water tests is presently being planned to prepare Newfoundlands for day-to-day life-saving jobs. They must learn not to hurt the person they are rescuing. They must also learn to handle a drowning person who is so scared that he could pull the dog underwater, too.

Newfs are now being trained to let a drowning person grab on to their tails or the hair on their backs just above their tails. The dogs swim up to their "victim" and circle to the left or to the right so the person can grab on to them.

Sebastian, a handsome white-and-black (Landseer) Newfoundland, turns to lick the face of his "victim" during a training session if he or she makes a crying sound. His owner, Elaine Lehr, trains him in the small lakes near their New Jersey home.

Water rescue training could be done in three to four months with a well-informed trainer and a willing dog, but it usually takes longer. Two years ago only a few dogs were being trained for water rescue, but now there are about a hundred dogs being trained in this country.

Clair Carr, of Michigan, is credited with starting water rescue work with Newfs. Her first Newfoundland puppy, Holly, was an exceptional dog that earned every obedience degree—the Water Dog Award for passing the Junior Division of water testing, and the Water Rescue Dog Award for passing the Senior Division.

Another exceptional dog was Ursa, who saved her owner's life in 1974.

Elizabeth Wiederhold had vacationed all summer on a small island off the coast of Maine with her poodles and Ursa. Ursa was a large, black Newfoundland who seemed to be more trouble than she was worth. She constantly shed, leaving drifts of hair wherever she lay down, and she often thought she was a lap dog, bounding up to her mistress and trying to climb up into her lap.

At the end of the summer season Mrs. Wiederhold made plans to take her little motorboat over to the mainland to pick up her husband and youngest daughter. They were going to help her close up the cottage for the winter.

Before leaving, Mrs. Wiederhold lighted several kerosene lanterns and put them in the windows. They would give a welcoming light for their return in the evening's darkness.

After an easy trip across to the mainland, Mrs. Wiederhold visited friends on their yacht while waiting for her family. When they didn't show up by 9:30 P.M. (a message phoned in by her husband saying they would not be coming that evening had not been delivered), she decided to return to the island. Although the air was now filled with a thick fog and heavy mist, Mrs. Wiederhold could only think of the lanterns burning away and her three dogs alone on the island.

In spite of her friends urging her to stay overnight, she headed out into the harbor, barely able to see past the end of her little boat. After leaving the harbor she still had a mile to go before reaching the island.

As she moved slowly forward with the aid of a small flashlight, Mrs. Wiederhold's fears began to grow. At last she realized that she had gone past the island and was heading out into the open ocean. Unable to see and not being able to follow the sounds of the foghorn, she did not know where she was. Then the little boat's motor stalled and died.

Mrs. Wiederhold was frantic. She began to yell for help. Who would hear her? Suddenly there was a bark. Mrs. Wiederhold recognized it as Ursa's and called her. Ursa swam up to the boat, appearing like magic out of the thick blackness and fog. The dog had heard the boat pass the island, recognized the sound of its motor, and knew something was wrong. She had jumped into the water and started swimming out to find her mistress.

Mrs. Wiederhold couldn't pull the dog into the boat, so she threw her a line and told her to take it home. Then she tried the motor again. This time it started up with a roar.

Ursa swam ahead of the boat while Mrs. Wiederhold followed along behind, shining the flashlight on her head so she wouldn't lose her. With an unbelievable sense of direction, the dog led her right to the boat's mooring. Ursa had saved her life! She could shed all she wanted to now, and there would be no complaints.

Ursa demonstrated the true qualities of a lifesaving Newfoundland. It is these qualities that many Newfoundland breeders hope can be used to save other lives.

In the future, it may become quite common to see a big, long-haired black dog standing on a beach alongside the lifeguard. It will mean double protection for swimmers and, it is hoped, a decrease in water accidents.

10
Donkeyball

"Gentlemen, start your engines and mount up," shouted the referee over the noise of the crowd. The basketball players tossed their reins over the donkeys' heads and scrambled onto their bony backs.

At the sound of the referee's whistle, eight little donkeys with their cumbersome riders leaning at various angles, plodded downcourt after the ball. Occasionally, they broke into a bumpy trot, but, often, after a few steps, they stopped short and ducked their heads. The wobbling riders didn't have a chance. Down the donkeys' necks they slid, tumbling to the floor.

Honey Pot, a stubby brown animal with a reputation for unloading her passengers in the time it takes to flick an ear, quickly pivoted away from her rider as he tried to remount. Frantically, he hopped after her, one leg thrown over her back. However, once he got on, the ride proved to be both rough and short. Honey Pot bucked furiously, each stiff-legged jump worse than the preceding one. Off he flew, landing in a heap on his backside.

"Here they come, folks—Back Breaker, Magnolia, Sleepy, and, of course, Honey Pot—ready to take on the challengers," the announcer called out.

The audience applauded and cheered as one of the helmeted players maneuvered his mount in closer to the basket, aimed, and shot. Missed! A teammate leaped off his donkey, retrieved the ball, remounted, and shot again. This time the ball caromed off the backboard and into the basket for two points. Red was ahead!

Now it was the White team's chance. Grabbing the

The White team moves downcourt at a walk while a lone Red player attempts to guard the basket

ball, their captain tucked it under his arm and urged Magnolia into a jolting trot. Resembling bumper cars in an amusement park, the rest of the players followed with toes dragging and fingers gripping the reins. Howls of laughter reverberated off the gymnasium walls.

Honey Pot was bringing up the rear now. Her rider had decided that walking was faster than riding, so he was pulling her along behind the others. That is, until she sat down and refused to budge. Tugging, name-calling, nothing could persuade her to move. She had had enough!

After what seemed, to the riders, like hours of hassling their donkeys from one end of the court to the other, the whistle blew. It was the end of the first half. Red was leading 10–6. Both teams' riders collapsed onto a bench to wipe their brows and rub their bruises.

Honey Pot stood quietly with the other donkeys, waiting patiently for the second half to begin. She didn't mind the games, or even her oversized riders, for she was always in command. It was just another night's work for her.

* * *

Donkeyball can be seen in school gymnasiums or on community baseball diamonds all over the country. Both donkey basketball and donkey baseball have tremendous audience appeal. They draw youngsters and adults alike to their games for an evening or afternoon of fun.

The donkey or burro players are hard workers, although the games are not usually long. The basketball games have eight-minute quarters, so even though the donkeys also participate in halftime activities, such as a game of Musical Burros, they are rarely out on the court for more than an hour and a half.

Donkey baseball uses ten donkeys—one for each of the players in the field, except for the pitcher and the catcher, and three who take turns being ridden by the players up at bat. A player must be on a donkey to catch or throw a ball or make an out, and any batter falling off between home plate and first base is out.

Honey Pot and most of her fellow competitors are of no particular breed. They range in color from white, gray, brown, or black to spotted and splotched variations of white and another color. They stand no more than 42 inches tall so their riders will be close to the ground with little chance of being hurt in a fall.

The smaller animals with one black stripe down the spine and another intersecting it at their withers and running over their shoulders in a perfect cross are called Sicilian donkeys. The cross is said to be a reminder that one of their ancestors carried Mary to Bethlehem on Christmas Eve. Today, these donkeys can still be seen carrying bunches of grapes down from the hills of Sicily to be made into wine that is shipped all over the world.

Everyone has his own style of riding

Honey Pot's riders end up on the floor more often than on her back, but many of the donkeys are quite rideable, with just a trace of stubbornness. None of them are mean, because if they bit or kicked intentionally, they could cause serious injuries both to the riders and to each other.

Many of the donkeys are big hams and actually seem to enjoy being in the spotlight of show business. They become restless and quite excited just before they are led out onto the basketball court or baseball diamond. Donkeys are quite smart—some say more intelligent than either horses or ponies—and will take advantage of their unsuspecting riders every chance they get.

Many of these long-eared competitors have developed their own stunts to unseat the players, while others are taught antics that cause audiences to break up with laughter. Kneeling, sitting, or lying down under the basket, shaking hands, putting their front feet up on a pedestal, giving kisses, darting off in the wrong direction, and, of course, bucking are some of the tricks guaranteed to shake up their two-legged teammates.

The donkeys work eight months a year. From November to the end of April is the basketball season, and from June to August is the baseball season. At least

Rigor Mortis watches her step while going down the truck's ramp

three different companies employ hundreds of donkeys to cover the country with their one-night games. Each of these little groups of entertainers includes ten donkeys and one driver-handler.

The same ten donkeys travel together all year, because after becoming used to one another, they work better together. Only eight donkeys participate in each game. The two extras are used as substitutes to give the regulars a day off.

The donkeys travel in trucks built specifically for them. Their driver is also the referee or umpire for the games. After arriving at each day's destination, the driver unloads his donkeys, waters them, and cleans out the truck. He also brushes and sometimes washes them, making sure to inspect their hoofs carefully so he can do any blacksmith work necessary before it becomes a serious problem. The donkeys wear soft, brown rubber shoes, instead of the iron shoes worn by horses. The rubber shoes protect the wooden floors of gymnasiums and also absorb some of the shock to the donkeys' legs that is caused by running on hard surfaces.

After the donkeys have stretched their legs, they are loaded back onto the truck. Sometimes there's a field nearby where they can grab a couple of mouthfuls of grass first.

The donkeys are usually fed a scoop of mixed feed—a combination of oats, corn, protein, and vitamins—twice a day. The servings are given in the morning, before they start the day's journey, and again in the evening after the game. The amount depends on the size of the individual donkey. The Sicilian donkeys, which may weigh around 100 pounds, don't require as much as one of the large animals that may weigh almost 400 pounds.

The donkeys' mangers are always full of hay so they can munch while traveling or snack between morning and afternoon naps. A well-cared-for and well-fed donkey will be full of pep during the games and put on a good show for the audience.

These cross-country travelers rarely become ill, because regular vaccinations and examinations are given by veterinarians along their route. The common cold is the one sickness that may occur in spite of preventive measures taken to guard against it. If a donkey does become sick, so sick that he is unable to travel, he is left behind with a veterinarian and picked up when he is well again. But this is hardly ever necessary.

The donkeys' sports equipment consists of a halter with a short lead rope attached to the rings on each side of the nosepiece to form reins. The riders must wear helmets to protect their heads.

Riders are assigned donkeys according to their size—smaller persons on the small donkeys and bigger people on the large donkeys. The smaller donkeys can carry up to 150 pounds, but their bigger companions can manage more than 250 pounds. Of course, too heavy a load is immediately and unceremoniously

An annual crop of young donkeys fills the ranks of the donkeyball players. This young donkey could someday become a donkeyball star

dumped on the floor without apology.

Riders are cautioned to be considerate of their donkey partners at all times. They must never jerk their heads around or pull their ears or tails. Any rider giving his mount a hard time and being overly rough is warned once. If he continues, he is thrown out of the game.

Donkeys who are treated cruelly would soon start kicking and biting to get back at their tormentors. This is too dangerous for everyone involved in the games, and even to bystanders watching the action, so any cruelty is not allowed.

Honey Pot and the other jennies (female donkeys) are bred to the jacks (male donkeys) and may give birth to a foal annually. The young donkeys are kept on the farms until they are big enough to be ridden, which is when they are about five years old.

Although these breeding programs contribute new donkeys to the teams every year, the handlers are constantly on the lookout for more donkeys, too. The biggest outside source is private families who decide to sell their donkey pets because the children have outgrown them or because they are moving to a smaller place and no longer have the room to keep them.

The two main requirements for a donkey to be considered for donkeyball are good health and an even temper. A donkey is allowed and often encouraged to be spirited during the games, but must be calm and gentle with young children, too. Most donkeys fit into this role quite easily because it is a natural part of their personalities.

In addition to allowing people to ride them, new donkeys must be taught to go up and down the truck's ramp for loading and unloading, walk in and out of school buildings, and climb up and down stairs. Among the most surefooted animals, donkeys usually need very little training to become used to these obstacles.

A newcomer is introduced to donkey basketball or

donkey baseball by being put right into a game and "playing" alongside his more experienced companions. It isn't long before he picks up the idea, and, soon after, the knack of getting his own way as much as possible.

Donkeys definitely have minds of their own and are often referred to as being stubborn. Donkey lovers disagree, though, and say they are just independent and more intelligent than most other animals. Donkeys are usually patient and gentle, but they can be quick to revolt if forced into something they don't want to do.

Honey Pot and her teammates may appear to compete against the team of Rigor Mortis, Nightmare, Droopy, and Dudley to the crowd's great amusement, but, after getting to know this little band of entertainers better, one begins to wonder who is really laughing at whom.

11
Aid Dogs

A blue station wagon pulled up to the curb in front of the school building. The driver, a tall, trim woman, got out, walked around to the passenger side, and opened the rear door. Out jumped a golden retriever.

The dog moved a couple of feet away from the car, turned and sat down. She looked eagerly at the teenager in the front seat.

"O.K., Cricket, I'm coming," the girl said to the dog.

Jill opened her own door and swung her legs out. The metal braces on them clanked together. Next she lifted her crutches out. Then Jill pushed herself up off the car seat and, balancing on her crutches, reached for a canvas bag containing her schoolbooks.

"Cricket, take it," commanded Jill.

Cricket grabbed the handle of the bag and waited patiently for Jill to shut the car door.

"Bye, Mom, see you after school."

"Have a nice day, dear," her mother answered.

"Cricket, heel."

Jill, who was born with cerebral palsy, started across

the sidewalk toward the school's front door. Her braces clanked and her crutches thumped with every step she took. Cricket walked happily beside her, head high and tail waving like a golden flag.

Other students were walking toward the school, too. None seemed surprised at the sight of the reddish-gold dog prancing alongside the young girl with long brown hair.

"Hi, Jill. Hi, Cricket," several students called out. Others waved at Jill, who nodded or smiled as she entered the school and moved slowly down the main hall. Cricket stayed right at Jill's side. She felt very important and knew that Jill depended on her.

They rounded a corner and started down another hall. The school was beginning to fill up with other boys and girls now, but Cricket didn't even glance at them. Jill continued on to her locker and stopped. Cricket sat down next to her and waited while Jill unlocked the door. After hanging up her jacket, Jill turned to Cricket and said, "O.K., Cricket, out."

Cricket opened her mouth and let Jill take the canvas bag from her.

"Thank you. Now, go find Mom."

Cricket trotted off down the hall. Jill turned to watch her until the dog rounded the corner and disappeared.

Cricket weaved between the students to the front door. She gave only a wag of her tail when they patted or called to her. She didn't stop, because she knew Jill's mother was waiting for her to return.

An Aid Dog learns to carry a crutch in the middle

In the afternoon, after Jill's last class, Cricket would again go to Jill's locker to carry her books back to the car. Since Jill's hands were occupied with her crutches, she couldn't always manage all her books and homework materials, too. During school she was given extra time between classes to return to her locker to get a book for the next class, but after school she usually had several books to carry home. Jill hated to ask a classmate to help her. She didn't like having her mother have to come into the school with her either. Cricket was a perfect assistant. She gave Jill the chance to be on her own without help from other people. Cricket was an Aid Dog.

* * *

Aid Dogs are dogs trained to help disabled people—people who cannot use their legs without crutches or braces and may even need a wheelchair to move around at all. It has been estimated that the number of disabled people in the United States may be as high as 38 million, with about 250,000 of them using wheelchairs.

An accident, a stroke, cerebral palsy, polio, and other diseases are the usual causes of paralysis or partial paralysis.

The story about Jill and Cricket is based on the real-life experiences of Doria, a golden retriever, and her owner, Sharon. Sharon is paralyzed in both legs as the result of an accident, and she has to use a wheelchair. Every morning, when Sharon was a high school student, Doria was at her side carrying her books inside to her school locker. Every afternoon Doria met Sharon at her locker and carried her books out to the parking lot, where Sharon's mother waited with the car.

Carrying books was only one of Doria's many jobs. At home she would fetch Sharon's brace, crutches, pencils, or books, and even carry a message to Sharon's mother. (Sharon would pin a note onto Doria's collar and tell her to take it to her mother.)

Doria is very important to Sharon, because Sharon doesn't like to have to ask others to do things for her. Doria, who has been trained to run errands for Sharon, helps to take the place of Sharon's legs. She knows how important she is to her mistress and enjoys helping her.

Having Doria at her feet or alongside her wheelchair gives Sharon a feeling of independence. For once, she is able to do things for herself without someone always waiting on her.

Doria became the first dog shown by an owner in an unmotorized wheelchair to earn a CD (Companion Dog) title in American Kennel Club licensed obedience trials. Sharon had to tie the lead to the chair because she needed to use her hands to push the chair. This did not bother Doria or Sharon, who both earned high scores every time they competed. At the end of their obedience exercises, Sharon and Doria were always applauded by the audiences.

Aid Dogs for the Handicapped Foundation is a nonprofit organization and operates on donations and dedication. Since its start in 1969, the Foundation has trained eight dogs. Seven are working for their owners in Pennsylvania, and one is in Charlotte, North Carolina.

Each of these dogs has been taught to fetch things for its owner and to give support if the person is in danger of losing his or her balance.

Any medium- or large-sized breed of dog may be trained as an Aid Dog. So far, all the dogs have been golden or Labrador retrievers. Most of them have been contributed to the Foundation by professional breeders. The retrievers seem best suited for this work be-

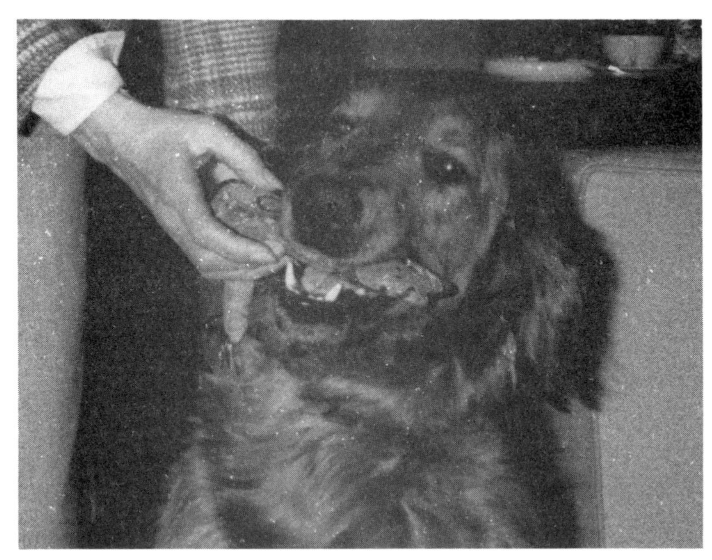

Golden retrievers have such soft mouths that they can carry glasses without breaking them

cause an Aid Dog must do a lot of fetching. Retrievers have been bred especially for fetching or retrieving, and, because of their soft mouths, hold things very gently. Newfoundlands may also be used as Aid Dogs. A 250-pound man who needed support when getting up or climbing stairs would have to have a larger dog, such as a Newfoundland, to help balance him.

Only spayed females are trained to be Aid Dogs, because of their gentle, easygoing manner. They must be in excellent health, have an even temperament, and show a sense of responsibility.

The puppies arrive at the ADH Foundation when they are between six and eight weeks of age. They are then put in foster homes until they are about a year and a half old. Here they receive plenty of love and attention while getting to know the world around them.

The puppies are taken on car rides, walked along city sidewalks, allowed to run through fields, taken on elevators, even introduced to swimming, if possible. They hear telephones ringing, stereos blaring, brakes squealing, and firecrackers banging. With careful handling, they also learn to trust people. All these experiences, which are impossible to find in the confines of a kennel, help to round out the puppies' personalities.

During these early months the puppies receive a little training, too. They learn how to come, to sit, to fetch a ball, and to walk on a leash. This first training is very relaxed and more of a fun time, because serious training will come later.

Foster families are carefully chosen by the Foundation. Usually the most eager foster parents are animal-loving youngsters ranging from ten to fifteen years of age. They have the time to work with the puppy and love doing it. Knowing that the puppy will grow up to help someone less fortunate than they are eases the pain of giving the dog away at the end of her puppyhood.

Before formal training begins, the puppy must be housebroken and out of the chewing-on-everything stage. When the puppy is about six months old, the foster parent starts her formal obedience training in a local obedience class. Mrs. Starr Hayes, president of the ADH Foundation, supervises the training.

Obedience training takes two to three months. The puppies learn to heel, sit, stay, lie down, stand, and come—all on and off lead.

After this training is completed, the puppy spends the next three to four months in custom training. However, until the basics of obedience training can be done without any mistakes, the custom training, which is more complicated, cannot be started. All training must

Starr Hayes demonstrates on Daffodil how an Aid Dog helps her handicapped owner to stand up

be built on a firm base so that it builds up gradually from easy commands to harder, trickier jobs.

The custom training of an Aid Dog depends on the new owner's needs. Once they learn the command "Fetch," the dogs can be trained to go and get eyeglasses, books, shoes, braces, newspapers, magazines, mail, pens, pencils, keys, or purses. They can be taught to carry small packages from the store, too.

When learning the command "Fetch," the dogs first start taking articles from the trainer's hand, then from a chair, then the floor, and then from another room. Finally, they are taught to bring the article to the trainer and hold it up to the person's hand so he or she doesn't have to bend down.

Aid Dogs who have to learn to pick up and carry a crutch must be taught to carry it in the middle. This way it will balance easier and not drag on the ground and jab the dog in the mouth. To train a dog to do this, tape is wrapped around the middle of the crutch and smeared with meat or gravy. When the command, "Fetch," is given, the dog will go right for the middle because of the meaty smell and taste. After she gets the idea to grab the crutch here, the meat can be eliminated and the tape taken off.

One of the biggest problems for a person on crutches is to pick up car keys, a comb, or lipstick that have dropped. They are difficult for a dog to pick up, too, because of their small size. Metal keys and key rings are especially annoying to dogs because they don't like the

Daffodil fetches her owner's purse and waits patiently for the command to give it to her

cold feeling of them in their mouths. However, Bit O'Honey of Golden Springs proves how talented an Aid Dog can be. Paul, her owner, says he never has to worry about picking up things. She has brought him articles ranging from a hairpin that she thought Paul might stumble over to a newly delivered, very heavy telephone book.

An Aid Dog can be trained to support her owner, too. If the person needs help in getting out of a chair or in getting up from the ground to a standing position, the Aid Dog will stand very still next to or in front of its owner. Then he can pull on the dog's collar, lean on her back just above the shoulders and pull himself up. Many disabled people lose their balance easily, so an Aid Dog also helps to pull and balance its owner when going up stairs.

During their training, Aid Dogs are taught to work in heavy traffic or on crowded city sidewalks, to ignore other animals, and to be courteous, but not too friendly, with other people. They are taught to become used to crutches and wheelchairs, too, although the training is not quite the same as working with a real disabled person.

The dogs learn to work with their owners in the final stage of their training. During this time both dogs and new owners get to know each other and learn how to work together.

Not every disabled person needs or is approved to receive an Aid Dog. Any person applying for an Aid Dog must be strong enough to control the dog and be willing to accept a dog into his home. Those who dislike or are afraid of dogs, are, of course, not chosen to receive an Aid Dog.

When interviewing someone applying for an Aid Dog, Mrs. Hayes makes sure the person knows what he or she has to put up with as the owner of a dog. She tells applicants that dogs shed hair, are sloppy when drinking water, may have an accident on the living room carpet, and will track dirt into the house if it is muddy outside. She tells them that Aid Dogs are living animals, not robots, and may make mistakes or need correction at times.

Mrs. Hayes also explains that Aid Dogs need exercise, have to be fed and brushed, and must be taken to a veterinarian for checkups and shots, or when ill.

The dogs have to be reminded of their obedience commands once in a while, too, so they won't forget them. If the person can't do this himself, then a neighbor, friend, or relative must take his place and put the dog through its paces.

A person who wants an Aid Dog must be able to afford the dog's care and have a proper place for the dog to sleep—not out on a porch or in a doghouse. Mrs. Hayes inspects the dog's new home before she recommends that the ADH Foundation board members give their final approval.

Most people who receive Aid Dogs are young—in their teens and twenties. Young people are not afraid to go out on their own with just a dog. They want to be independent and are usually more willing than older people to try to work with an Aid Dog. Usually, a younger person's muscles are stronger, too, and a younger person will accept a dog's needs more easily than an older person who is unused to animals.

New owners take care of their own dogs as much as they are physically able. They feed, brush, and exercise their dogs themselves. This gives them a sense of responsibility and a feeling that a living creature depends on them. Often they will think, If I can do this, I can do other things, too.

Aid Dogs give disabled people a chance to communicate with another living being. Paul, owner of Bit O'Honey, says, "Now I have a reason to get up in the morning."

And indeed he does. Honey needs to be fed and let out every morning. Paul leashes her inside the house and then moves to the porch on his crutches to hook her to a backyard exercise line. Honey is Paul's partner, and taking good care of her is his way of thanking her.

Once a person is qualified to receive an Aid Dog, he is matched up with a dog. A mild, gentle person will receive a quiet dog. A stronger, more forceful person will be given a more active, energetic dog.

During the time that the dog and the new owner are getting to know each other, Mrs. Hayes allows the dog to visit her new home-to-be. A day or two before the Aid Dog is to go permanently with her new owner, Mrs. Hayes leaves her in a boarding kennel. When the dog is picked up, it is by her new owner and off she goes to her new home. Most dogs are so happy to leave a boarding kennel that any familiar face is a joy to them. They are so excited that they don't even seem to miss their old home.

One man was so proud of his dog's intelligence that when he got her home he had to show her off to some friends.

"Fetch pillow," he commanded. Off the dog went to fetch a pillow. In fact, she fetched all the pillows she could find, and her surprised owner finally had to call Mrs. Hayes to find out how to stop her.

The ADH Foundation tries to check up on the dog and her new home a few months after placement to make sure everything is all right. It doesn't turn over the dog's registration papers until after a six months' trial period. This is the Foundation's way of keeping control of the dog in case she isn't working out or isn't being cared for properly.

Aid Dogs learn to adjust to their owners' needs. They become very aware of their problems and will even go get something before the command is given, such as fetching a pair of glasses when the owner picks up a book to read. Doria learned to tuck her body under Sharon's wheelchair whenever they went through a doorway so she didn't get caught between the chair and doorframe while walking in the heeling position.

In the future, the Aid Dogs for the Handicapped Foundation hopes to have its own training center with a house, kennel, and full-time trainer. A new owner could come to live there while he is getting to know his dog.

Disabled people will be hired to do the secretarial work, help with the dogs, and do other jobs around the center. This will give the young dogs a chance to get used to these people and their aids—crutches, braces, wheelchairs, walkers, etc. It would also help those who have come to receive their dogs to see what other disabled people have accomplished.

In addition to supporting her owner and fetching various articles, an Aid Dog is a companion and a friend. True, she will fetch a Kleenex (slightly damp) when no one else is around and the box is out of reach, but she also gives her owner a reason to get out, meet other people, and become involved in outside activities.

Before Doria came to live with her, Sharon had no reason and no desire to leave the house. Without Doria, Sharon would never have become so involved in competing at dog shows. Having to walk Doria every day was Sharon's first step in getting to know the outside world again.

Aid Dogs and their owners together build a need for each other based on love and respect, not sympathy. More puppies and their owners-to-be are getting ready for each other every day.

When asked why he wanted an Aid Dog, Keith, who had had a stroke at age twenty that left him without control of the left side of his body, said, "Because a wet nose is love."

12
The Rasslin' Bear

"Ladies and gentlemen, today's star of the show—Victor, the Rasslin' Bear!" boomed the announcer over the loudspeaker.

"He stands eight feet three inches tall, weighs 650 pounds, and has a winning record of 50,000 to one."

The crowd hummed with excitement as trainer George Allen, dressed in a purple one-piece bodysuit and a shimmering purple overblouse, led Victor onstage. The huge brown bear ambled over to a corner of the blue wrestling mat and sat down, with hardly a glance at the audience.

Victor's wrestling "uniform" included a strong leather muzzle and an unbreakable collar to which a chain lead was attached. Swinging his large head around, Victor wiggled the end of his nose, trying to catch the smell of marshmallows. He loved marshmallows, cookies, Kool-Aid, candy, popcorn, and other goodies, and licked his chops at the thought of how good one of them would taste right now.

Waiting a minute for the audience to quiet down,

the announcer continued: "Today we have four men who have volunteered to wrestle Victor. They are going to try to put him on his back. The first challenger is Bill . . ."

The noise from the crowd drowned out the rest of the announcement. All at once everyone stopped talking and watched a large muscular man in jeans, T-shirt, and sneakers walk to the center of the mat. He turned and faced the gentle-looking bear, who was now lying down on his belly with his head resting on his front paws.

"Come on, Victor, get up," said George as he urged Victor to his feet and unsnapped the lead. The bear pushed himself up and looked at his challenger, who nervously watched his every movement.

"Let the match begin," said the referee. He blew his whistle.

"O.K., Victor, get him!" commanded George as he gave the bear a shove.

Victor lumbered forward, head swinging back and forth, massive shoulders rolling with each step. He closed in on Bill, who stood crouching slightly with feet apart and arms spread. Bill began to wish he was anywhere·but in the middle of that mat. How did he get himself into this?

Just before Victor reached Bill, he stood up to his full height and walked the last few steps on his hind legs. Throwing his front legs around the man, Victor enveloped him in a bear hug, burying Bill's face in the

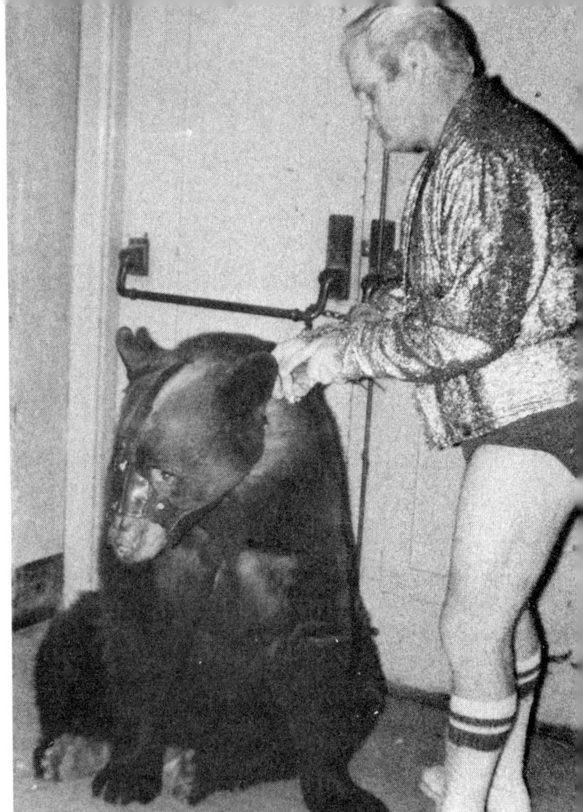

A last-minute brushing and Victor is ready to meet his challenger. Trainer George Allen puts on Victor's muzzle

thick, shaggy fur of his chest. Bill struggled to keep his balance as the two "danced" around the platform. Then Victor began to lean forward, throwing his weight onto Bill's shoulders.

Trying to avoid falling on his back, Bill turned and collapsed to his hands and knees. Victor was all over him, pushing him first to one side and then the other. Bill tried to crawl away from the bear, but Victor grabbed one of his legs with a hooked right front paw and pulled him back.

"Easy, Victor, easy," cautioned the trainer. He knew

the bear must not use all his strength in his powerful front legs. He could easily break the man's ribs.

Flipping Bill over on his back, Victor pressed him into the mat with his chest. Bill twisted and strained, trying to push the bear off and roll over onto his stomach, but he didn't have a chance. Just as the referee slapped the mat, signaling a pin, Victor licked the side of Bill's face with his long wet tongue. Sympathy? Hunger? Bill wasn't sure which, but he did know he wanted OUT of the bear's grip—and right now!

George hauled the bear off the exhausted man, who lay gasping for breath.

"Come on, Victor. Let him up now."

Victor backed away and casually strolled over to his corner, where George gave him a couple of marshmallows as a reward. Victor slurped them down in one gulp and looked around for his Kool-Aid. George handed him a bottle half filled with the sweet liquid. Victor sat back on his haunches, grasped the bottle in his front paws, and poured the orange drink down his throat.

The next challenger was nervously pacing back and forth on the other side of the platform, waiting for the referee's signal to approach the center of the mat. But he didn't last long either. Victor made short work of him and his next two opponents, too.

Four pins in less than five minutes isn't bad for one wrestler. And Victor does it at almost every show every day. This is Victor's job—to wrestle and pin all challengers. There are many people who want to brag about how they wrestled a bear to the ground. From professional athletes to movie actors to average fathers, husbands, and boyfriends—the list of challengers is endless.

* * *

Victor, an Alaskan brown bear, is fifteen years old. He was born at Grant's Farm near St. Louis, Missouri. When he was only three weeks old, his eyes not quite open yet, his mother died. The orphan, a little bundle

Victor pins this opponent in less than 60 seconds

of thick, soft brown fur, was adopted by George Allen, who used to be a professional wrestler himself.

George raised Victor just as a mother would her baby. He fed him milk from a bottle, baby cereal, and chunks of apples and carrots. He even carried him around in his pocket everywhere he went.

Today, Victor, who is owned by Victor Promotions of Cherokee, North Carolina, eats forty pounds of food daily. He has two meals and numerous snacks of Purina dog chow, lettuce, Kool-Aid, cookies, candy, carrots, celery, apples, and his favorite, marshmallows. George always keeps a trunkful of Kool-Aid, cookies, and marshmallows at ringside to reward Victor before, during, and after each show. Victor's grocery bill is $77 a day.

When Victor was three months old, George trained him to walk on a leash. Victor fought it at first—backing away and straining against the pull of his collar, but George patiently worked with him every day. Finally, Victor settled down and learned to walk quietly at George's side.

Next Victor was trained to wear a muzzle. He tried to rub and pull it off his head at first. George would give him a bottle of Kool-Aid before it was put on to encourage him to wear it. A muzzle is necessary for safety reasons. If Victor ever gets excited or too fierce during a match, the muzzle keeps him from biting.

Bears have long, heavy claws, five on each foot, which can rip an enemy apart. Victor's claws were removed when he was a cub, because otherwise he would be too dangerous to wrestle. His front teeth and four long canine teeth have also been removed for safety reasons, but he has thirty-two molars left for grinding and chewing his food.

Although Victor resembles a large, soft, brown Teddy bear—one of those cuddly stuffed animals that sits on the end of a bed, he is definitely not a toy or a family pet. George is very cautious when working with Victor and makes certain that other people don't get too close to him. Even during a wrestling match, George is only a few feet away from Victor, in case the bear gets overly aggressive with his challenger.

George taught Victor several different wrestling holds—single leg takedown, ankle pick, body press, and others—to use in a match. Although bears are natural-born wrestlers and roll around on the ground and scuffle as cubs, it took two years for George to teach Victor the various holds and, more important, to learn not to use all his strength. One blow from the front paw of an adult grizzly, polar, or brown bear can kill a man.

During his training Victor was never hit or beaten, because an animal that has been beaten is untrustworthy and may suddenly and without warning attack his trainer or anyone else within reach. A happy, confident bear who is rewarded rather than punished is a better worker. A rough "No," and not getting his candy or Kool-Aid, is Victor's only punishment.

Victor has performed at sports shows and other

events in all the forty-eight continental states plus Hawaii, and in Canada and Puerto Rico. He has flown in airplanes, but usually travels in either a van or a 72-passenger school bus. The bus has had most of the seats removed so it can easily hold Victor, his equipment and food, and a portable stage. Victor sleeps between shows, but the motion of the bus usually keeps him awake when traveling.

To protect Victor in the various states he must have regular examinations and vaccinations from veterinarians to keep him healthy.

Victor's home is a large aluminum cage with barred windows on each side and barred doors at each end. He sleeps on a bed of straw on top of a layer of sawdust. A pail of drinking water cannot be left in his cage because Victor immediately tips it over and splashes around in it.

Victor cleans himself, but he also gets brushed every day and combed before each show. His thick fur is very soft, and he enjoys having his back scratched when being groomed.

Bears have small eyes and can't see well. They can't hear very well either, but they do have an excellent sense of smell. Bears don't perspire, so they have to pant just like dogs to cool their bodies when hot.

Victor is not the only bear that wrestles people. The original Victor was a Canadian black bear who started touring the country in 1962.

Today, Victor is joined by at least two other bears

Victor gratefully accepts a marshmallow after his performance

accepting challenges from two-legged wrestlers. A black bear, also named Victor, who stands seven feet tall and weighs 500 pounds, is owned by the same company. He and Terrible Toby, another black bear who weighs 600 pounds, are also touring the country looking for people to wrestle.

People stand in line to wrestle the bears and are very disappointed when the daily quota is filled. In order to wrestle Victor, challengers must agree to the Bear Rasslin' Rules. They include not kicking, not touching Victor's muzzle or collar, and being at least eighteen years old. George does not allow anyone to slap or sock Victor either and will stop the match immediately if anyone tries to hurt him.

Most challengers try to muscle Victor, using their strength against his. They don't get anywhere. A champion arm wrestler from Connecticut and a professional wrestler from California have given Victor tough matches, but didn't pin him. It was a young former high school wrestler from Pennsylvania (weighing about 130 pounds) who found out that it takes actual wrestling holds to make Victor lose his balance so he can be taken down. He also discovered that it takes more than wrestling moves to *keep* him down.

Victor loves to wrestle. It's just a game and a lot of fun for him. But then again, it's always fun when you are always the winner.

Index